Chemistry and Action of Herbicide Antidotes

Edited by

Ferenc M. Pallos
Western Research Centers
Stauffer Chemical Company
Richmond, California

John E. Casida
Department of Entomological Sciences
University of California
Berkeley, California

ACADEMIC PRESS   New York   San Francisco   London   1978
A Subsidiary of Harcourt Brace Jovanovich, Publishers

ACADEMIC PRESS RAPID MANUSCRIPT REPRODUCTION

ACADEMIC PRESS, INC.
111 Fifth Avenue, New York, New York 10003

United Kingdom Edition published by
ACADEMIC PRESS, INC. (LONDON) LTD.
24/28 Oval Road, London NW1   7DX

Library of Congress Cataloging in Publication Data

Main entry under title:

Chemistry and action of herbicide antidotes.

"Presented at a symposium . . . at the 173rd
national meeting of the American Chemical Society
held in New Orleans, Louisiana on March 24, 1977."
   1.   Herbicide antidotes (Plant protection)—
Congresses.   I.   Pallos, Ferenc M.   II.   Casida,
John E.   III.   American Chemical Society.
SB951.45.C48   632'.954   77-25191
ISBN 0-12-544050-2

Contents

List of Contributors

Numbers in parentheses refer to the pages on which authors' contributions begin.

Duane R. Arneklev (15), Mountain View Research Center, Stauffer Chemical Company, Mountain View, California

A. S. Bhagsari (21), Department of Chemistry, Fort Valley State College, Fort Valley, Georgia

Mervin E. Brokke (15), Stauffer Chemical Company, Eastern Research Center, Dobbs Ferry, New York

John E. Casida (151, 161), Laboratory of Pesticide Chemistry and Toxicology, Department of Entomological Sciences, University of California, Berkeley, California

F. Y. Chang (35), Department of Environmental Biology, University of Guelph, Guelph, Ontario, Canada

D. Stuart Frear (133), Metabolism and Radiation Research Laboratory, Agricultural Research Service, U.S. Department of Agriculture, Fargo, North Dakota

Reed A. Gray (15, 67, 109), Mountain View Research Center, Stauffer Chemical Company, Mountain View, California

Otto L. Hoffmann (1), Gulf Oil Chemicals Company, Merriam, Kansas

Robert M. Hollingworth (109), Department of Entomology, Purdue University, Lafayette, Indiana

Grant K. Joo (67), Mountain View Research Center, Stauffer Chemical Company, Mountain View, California

Ming-Muh Lay (151), Laboratory of Pesticide Chemistry and Toxicology, Department of Entomological Sciences, University of California, Berkeley, California

Gerald L. Lamoureux (133), Metabolism and Radiation Research Laboratory, Agricultural Research Service, U.S. Department of Agriculture, Fargo, North Dakota

J. Bart Miaullis (109), Mountain View Research Center, Stauffer Chemical Company, Mountain View, California

John J. Murphy (109), Mountain View Research Center, Stauffer Chemical Company, Mountain View, California

Ferenc M. Pallos (15), Stauffer Chemical Company, Western Research Centers, Richmond, California

B. A. Phillips (21), Department of Chemistry, Fort Valley State College, Fort Valley, Georgia

Richard H. Shimabukuro (133), Metabolism and Radiation Research Laboratory, Agricultural Research Service, U.S. Department of Agriculture, Fargo, North Dakota

F. W. Slife (63), Department of Agronomy, University of Illinois, Urbana, Illinois

G. R. Stephenson (35), Department of Environmental Biology, University of Guelph, Guelph, Ontario, Canada

Victor M. Thomas (109), Mountain View Research Center, Stauffer Chemical Company, Mountain View, California

Robert E. Wilkinson (85), Department of Agronomy, The University of Georgia Agricultural Experiment Stations, Experiment, Georgia

Preface

Chemical herbicides play a major role in promoting high crop yields by reducing competition from undesired weed species for available nutrients, water, light, and other essential environmental factors. They also replace more expensive labor and mechanical methods for weed control. For maximum utility in crop production, herbicides must have a high degree of selective toxicity, i.e., injury to undesirable plant species but not to crops.

Selectivity may be governed by one or more factors including herbicide penetration, uptake, translocation, and metabolism. For example, corn plants have an inherent ability to detoxify chloro-s-triazine herbicides at a rate that far exceeds the detoxication potential of certain weed species associated with corn culture. There are other groups of chemical herbicides that have desirable attributes in weed control but do not possess in all instances a sufficient margin of selectivity. An added degree of selectivity can often be achieved by critical timing of application, suitable placement of the herbicide, or now by a third approach with chemical safening agents, the subject of this book.

The term "herbicide antidotes" may or may not be the best description of these chemical agents. "Herbicide antagonists," "crop protectants," and "herbicide safeners" may be just as appropriate and are indeed used interchangeably in this developing field. The first practical breakthrough with chemical safeners was achieved by Otto Hoffmann of Gulf Oil Chemicals Company, the "father of herbicide antidotes." He found that 1,8-naphthalic anhydride is active as a seed treatment antidote for thiocarbamate herbicide injury in corn. This development is useful only under restricted conditions involving prior treatment of crop seeds with the safener. The next historical highlight in the development of herbicide safeners was the discovery of the action of dichloroacetamides by Stauffer Chemical Company researchers. These compounds provide a high measure of protection to corn plants from injury associated with the use of thiocarbamate herbicides. Furthermore, most of these compounds do not require treatment of crop seeds but can be jointly formulated or applied as a tank mix.

Several investigators have developed evidence to pinpoint significant facets of the mode of action of herbicide safeners. Studies of this type lay the fundamental background for improvements in antidote design and use.

If this emerging field of herbicide antidotes leads to better applications of modern herbicides and to improvements in crop production and yields, it is well worth the effort of the many researchers active in this area. The editors hope and anticipate that this book will contribute to an understanding of the chemical and biochemical basis for the further development of herbicide antidotes. The contributions contained herein were presented at a symposium under the same title at the 173rd National Meeting of the American Chemical Society held in New Orleans, Louisiana, on March 24, 1977. The editors express special thanks to Julius J. Menn for suggestions and encouragement in organizing this symposium.

We share the convictions expressed in the following quote from André and Jean Mayer [*Daedalus* **103**, 83 (1974)]:

> Few scientists think of agriculture as the chief, or the model, science. Many, indeed, do not consider it a science at all. Yet it was the first science—the mother of sciences; it remains the science that makes human life possible; and it may well be that, before the century is over, the success or failure of science as a whole will be judged by the success or failure of agriculture.

Part I

Introduction

HERBICIDE ANTIDOTES: FROM CONCEPT TO PRACTICE

Otto L. Hoffmann
Gulf Oil Chemicals Company

Herbicide antidotes in the past few years have gone from experimental curiosities to practical realities. Initial observations that led to the commercial products were 2,4,6-trichlorophenoxyacetic acid (2,4,6-T) antagonism of 2,4-D on tomato and the total inactivation of barban by 2,4-D. The development of a fast, cheap and accurate detection method for herbicide antidotes led to the development of 1,8-naphthalic anhydride as an antidote for thiocarbamate, chloroacetanilide and dithiocarbamate herbicides on corn. In addition to these families of chemicals 1,8-naphthalic anhydride also has utility for antidoting alachlor on sorghum and rice, molinate on rice and barban on tame oats. Besides naphthalic anhydride some halogenated alkanoic acid amide-type chemicals have shown antidoting properties against several herbicides.

I.   INTRODUCTION

The picture in Figure 1, shown at the 1969 meeting of the Weed Science Society of America, illustrates total antidoting of a lethal rate of EPTC (S-ethyl dipropylthiocarbamate) on corn with 1,8-naphthalic anhydride (1). Only one ounce per acre (70 g/ha) of this antidote was needed to counteract the effect of nearly 100 times greater quantity of EPTC. For the first time, it was possible to chemically control unwanted plant growth while permitting the growth of an economic crop of the same genus, species and variety. This finding served as an impetus for several organizations to initiate research on antidote-type chemicals. Once this information was available, it appeared so simple, but getting to this result entailed many trials over the time span of 21 years.

Fig. 1. *Corn antidoted against 6 lb/A EPTC by dusting seed before planting with 0.5% by weight 1,8-naphthalic anhydride. Untreated corn seed planted on right row.*

II. PHENOXYACETIC ACIDS AND OTHER GROWTH REGULATORS AS ANTIDOTES

My first interest in herbicide antidotes was aroused in 1947 in a greenhouse filled with tomato plants that were sprayed with 2,4-D (2,4-dichlorophenoxyacetic acid) analogs. The vents had not been opened until well into the afternoon of a hot summer day. At first glance, all of the plants, including the untreated controls, appeared to be dying from 2,4-D fumes. However, closer inspection showed plants that had been treated with 2,4,6-T (2,4,6-trichlorophenoxyacetic acid) were apparently normal. This antagonistic relationship between 2,4,6-T and 2,4-D is illustrated in Figure 2 (2,3).

2,4,6-T was found to antagonize all types of rapid growth, such as cell elongation and multiplication on tomatoes, and seemed to fit into what was then a popular explanation of biological chemical interactions - the lock and key mechanism. It did not antagonize a lethal rate of 2,4-D. So do tomato plants "grow to death?" The use of antidotes

Fig. 2.   2,4-D antidoted with 2,4,6-T on tomatoes.
Left to right: check; 2,4-D 10 ppm; 2,4-D 10 ppm, 2,4,6-T
100 ppm; 2,4,6-T 1000 ppm (3).

as a research tool had been found to be productive in
vitamin research (4) since the vitamin, p-aminobenzoic acid,
was discovered through its antagonism of the bactericide,
sulfanilamide.

The lock and key explanation of antidote action did not
apply to the total antidoting of oats of another herbicide,
barban (4-chloro-2-butynyl-m-chlorocarbanilate) by 2,4-D (5).
Structures of 2,4-D and barban are quite dissimilar and they
do have different growth actions which could possibly explain
their antagonism. 2,4-D accelerates plant growth whereas
barban slows down growth. Both would fit into the growth
regulatory category in Overbeek's division of herbicides (6).

An effort was made to find a practical use for the
effect between barban and 2,4-D. For this purpose, seed
treatments would be necessary because foliar spray of barban
and an antidote would antagonize both the wild oat pest and
the wheat crop. However, 2,4-D, 2,4,6-T and MCPA (2-methyl-
4-chlorophenoxyacetic acid), all of which antidoted barban as
foliar sprays, were too toxic to be useful as seed treatments.
From screening trials, chemicals had been found which caused

deformed foliage on tomatoes,as did 2,4-D, 2,4,6-T and MCPA, but which were safe as dressings on grass crop seed. Many of the growth regulators when tested by applying to wheat seed proved to be barban antidotes and prevented injury from high rates on wheat. This finding is illustrated in Figure 3 by reduction of wheat injury, when the seed was treated before planting with 1 oz/bu (1 g/kg) of 4'-chloro-2-hydroxyimino acetanilide (7-9).

Fig. 3. Barban antidoted with 4'-chloro-2-hydroxyimino acetanilide on wheat. Left to right: 2 lb/A barban, 1 oz/bu antidote; 2 lb/A barban, no antidote; 1 lb/A barban, 1 oz/bu antidote; 1 lb/A barban, no antidote. In each pot wheat is on the left side and oats on the right.

III.   NEED FOR ANTIDOTES

In considering a systematic approach to antidote research several factors that could contribute to a successful outcome were investigated. Let us look first at the need for antidotes.

If we accept the desirability of chemical weed control in place of cultural weed control, then it is apparent that if even one weed species escapes the applied herbicide, at least some mechanical control will be necessary. It is doubtful if we will ever find a chemical that by itself will selectively control the following problems: weedy corn in a seed corn field, shattercane in a grain sorghum field, red rice in a rice field, wild oats in a tame oat field, weedy millets in a millet field, or wild beets in a sugar beet field. Even now, after some 40 years or more of chemical weed control, there are no totally satisfactory broadleafed herbicides for broadleaf crops; no totally satisfactory grass herbicides for grass crops.

IV. CONSIDERATIONS IN SCREENING METHODS FOR ANTIDOTE
     DETECTION

     For progress in antidote research an economical detection method was desirable. For screening procedures we need to consider the crops, the herbicides and operation mechanics.

     To test an antidote on 100 herbicides and 50 annual crops gives a potential of 5000 observations needed to completely evaluate each candidate at one rate. But an antidote rate may fail because it is either so high it causes crop injury or so low that it is ineffective. With only two antidote rates to measure activity, a complete evaluation of a candidate antidote on 100 herbicides would require 10,000 items of data. Therefore, to be practical both the herbicides and the crops for screening would have to be selected.

     Overbeek's (6) classification of herbicides was useful for selecting candidate compounds as he divided herbicides into growth regulators and photosynthetic inhibitors. These two groups can be divided on the basis of antidote response as shown in the following scheme.

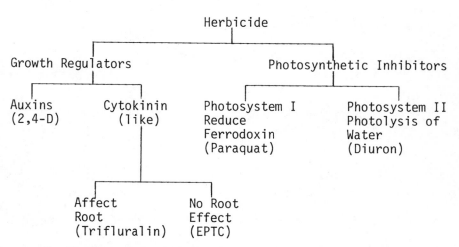

     In the growth regulator group are auxins, represented by 2,4-D, as determined by the tomato response and cytokinin-

like (10) herbicides, which are recognized by the oat response. The cytokinins can be further divided: those that
affect root growth as does trifluralin (α,α,α-trifluoro-2,6-
dinitro-$\underline{N}$,$\underline{N}$-dipropyl-$\underline{p}$-toluidine) and those that don't affect
root growth as represented by EPTC. The group of photo-
synthetic inhibitors can be divided into those that affect
photosystem I,as represented by paraquat (1,1'-dimethyl-4,4'-
bipyridinium dichloride), and those that affect photosystem
II as represented by diuron [3-(3,4-dichlorophenyl)-1,1-
dimethylurea]. A representative herbicide from each group
would have to serve in an initial screen. As representatives
of the different groups, five herbicides were chosen: 2,4-D,
EPTC, trifluralin, paraquat and diuron. Four of these groups
can be antidoted to some extent on some crop. The antidoting of 2,4-D has already been mentioned. Trifluralin can be
antidoted on corn in petri plate tests. However, soil tests
did not lead to acceptable results. Paraquat can be antidoted with oxidation-reduction compounds, when both paraquat
and the antidote are applied as sprays. Paraquat at 0.25
oz/A (18 g/ha) can be antidoted with ferrous sulfate at 2 lb/
A (2240 g/ha) (Figure 4). Seed treatment of wheat with
ferrous sulfate gave only a minimal antidoting response.

*Fig. 4. Paraquat antidoted with ferrous sulfate on
wheat and oats. Left to right: check; paraquat 0.25 oz/A;
paraquat 0.25 oz/A, ferrous sulfate 2 lb/A.*

No success was achieved in antidoting diuron or other herbicides affecting photosystem II. Over 6,000 chemicals were tested against this type of herbicide with no detectable antidoting. Figure 5 illustrates one test of over 2,000 candidate antidotes with an experimental urea herbicide.

Fig. 5. No antidoting of Hill reaction inhibitors was achieved on over 6,000 tests. In this picture of over 2,000 antidote candidates, corn check plants stand out boldly.

V.   CORN-EPTC SCREEN FOR ANTIDOTES

EPTC was selected as a candidate herbicide for the group of cytokinins not affecting root growth. From previous work (7) it was known that EPTC could be antidoted on corn.

The procedure for screening antidotes on EPTC was as follows: to 5 g corn seed in a 3-dram vial (Wheaton is best) were added 150 mg (3%) or 25 mg (0.25%) chemical and 0.050 ml (1%) methanol. The seed was shaken in a Spex mixer grinder for 20 seconds. The methanol acted as a grinding aid and sticker for the chemical. The seed acted as grinding pellets. With over 99% of the chemicals tested, size reduction and adherence at 3% was excellent. Only a few viscous tars gave poor coverage with this treatment method.

Five seeds of each of the treatments were planted in loam soil in flats 1.5 in (3.8 cm) from five untreated seeds

Fig. 6. Corn seed planted for antidote assay. The unstaked check rows provided a close by measure of effectiveness of each antidote and also provide a means of detecting those antidotes that move in the soil.

(Figure 6). In effect, this superimposed a large degree of uniformity in the tests. Tests were rated 3 weeks after planting with a 3-digit code. The first digit gave the number of plants emerged, the second digit the number of plants distorted by EPTC, and the third digit the height of the plants divided into five categories. A rating of 505 indicated 5 plants emerged, none distorted by EPTC and a height 75% or more of untreated control height.

The rating system classified antidotes in 101 categories. This permitted a total ranking of the active chemicals from trace activity to candidates for field testing in just one test with five treated seeds. Ease and economy are advantages of the seed treatment assay. In addition, potentially effective soil applied antidotes can be detected with this assay. This is seen when the check row is antidoted by the chemical moving from the treated row to the check row.

VI.   MULTIPLE CROP-MULTIPLE HERBICIDE SCREEN FOR ANTIDOTES

The best antidotes were tested further in a multiple crop-multiple herbicide screen. To aid in this assay, a seed selector was designed that permitted the selection of a small quantity of corn, oat, rye, rice, barley, wheat, sorghum, cotton, pea, soybean, radish, sugar beet, alfalfa, tomato and flax seed. This seed was placed in a 3-dram vial, treated with the appropriate amount of chemical and 1% methanol. It was shaken 20 seconds on a Spex mixer and planted in a 6 in x 12 in (15 cm x 30 cm) flat. The herbicide test chemical was applied directly over the seed. The seed was covered with soil and the flat was watered. Plant growth was rated in comparison to a treated and untreated check.

These methods led to the discovery of the first commercial herbicide antidote (11) (Figure 7) and many others (12) (Figure 8).

Fig. 7. EPTC (8 lb/A) antidoted with 1,8-naphthalic anhydride seed treatment. Rates of 1,8-naphthalic anhydride, % w/w on seed: left to right; 2%, 0, 1%, 0.5%, 0.25%.

To increase the efficiency of assaying soil applied antidotes, a multiple crop flat planter was made that plants at one time 14 crops. The crops that are planted can be altered by changing planter valve bodies. Flax, sugar beets, corn, millet, wheat, oats, barley, sorghum, soybeans, cotton, rice, cucumber, radish and alfalfa are usually planted. The herbicide and antidotes are applied directly over the seed, covered and watered. Rating of growth was in comparison to

Fig. 8. Screening test of halogenated alkanoic acid amides.

a herbicide treated and untreated check.

Approximately 40% of all chemicals tested by the corn seed treatment method antidoted EPTC to some extent. Over 1,600 chemicals were found to be highly effective with ratings of 414 or better. About 1,400 of these 1,600 antidotes are amides or compounds such as ketones, acids and amines.

The multiple crop-multiple herbicide assays showed that practically all grassy crop plants can be antidoted to some extent by one or more of the cytokinin herbicides. Antidoting is a very specific crop/herbicide/antidote interaction. Corn can be antidoted against most of the cytokinin herbicides that do not affect roots. Alachlor [2-chloro-2',6'-diethyl-N-(methoxymethyl)acetanilide] can be antidoted on many grass crops including rice, grain sorghum, wheat, oats, barley and rye, by more antidotes than any other available herbicide.

VII.   1,8-NAPHTHALIC ANHYDRIDE AS A SEED TREATMENT ANTIDOTE

Of all the chemicals found, 1,8-naphthalic anhydride is the most versatile of the seed treatment antidotes (13). It has antidoted corn against all of the thiocarbamate, dithiocarbamate, and chloroacetanilide herbicides on which it

was tried. It is useful on rice and grain sorghum to antidote alachlor, on rice against molinate (S-ethyl hexahydro-1H-
azepine-1-carbothioate), on tame oats against barban, and has
some activity on other cytokinin herbicide-grass systems.

VIII.   ANTIDOTES IN RELATION TO DORMANCY

      Another interesting effect of 1,8-naphthalic anhy-
dride found in testing was on cis Δ 4-tetrahydrophthalimide
and potassium nitrate on sugar beets. These two chemicals
have been described as germination inhibitors present in sugar
beet seed balls (14, 15). When sugar beet seed was placed in
contact with these chemicals, it failed to germinate. If
seed was first treated with 1,8-naphthalic anhydride, it
would germinate in the presence of these two inhibitors.
Potassium nitrate is particularly interesting in that it
promotes dormancy of sugar beets and breaks dormancy of wild
oats (16). The same chemical causes effects in opposite
directions - dormancy and dormancy breaking.

      Could this be true of cytokinins? There are literature
references on breaking of seed dormancy with kinetin (17).
Many of the cytokinin herbicides break dormancy of weed seeds
(18). Could the effect of EPTC, barban and similar compounds
be a promotion of bud dormancy in affected plants? If so,
then the introduction of a bud dormancy breaking chemical
should reverse the effects of chemicals like EPTC. Ether
(19), 2-chloroethanol (20) and 1,2-dibromoethane (21) are
described as breaking bud dormancy of various plants. With
gas confining testing procedures where EPTC treated flats
seeded with corn were held in polyethylene bags, it was
demonstrated that these three chemicals are indeed antidotes
of EPTC on corn. Of these, 1,2-dibromoethane is the most
active (Figure 9) and ether is the weakest. 2-Chloroethanol
has a narrow margin between phytotoxicity and effectiveness.
This work suggests that one of the mechanisms of toxicity of
the cytokinin herbicides is through induction of precocious
bud dormancy in seedlings. The antidotes may be preventing
this dormancy.

      The potential of herbicide antidotes is very promising,
both for increasing the utility of herbicides and for helping
in the elucidation of herbicide action mechanisms.

Fig. 9. EPTC antidoted with 1,2-dibromoethane. Left 8 lb/A EPTC; right 8 lb/A EPTC + 1,2-dibromoethane.

IX.   REFERENCES

1. Hoffmann, O. L., Weed Sci. Soc. of Amer. Abstracts 12 (1969).
2. Hoffmann, O. L., "Some Effects of Plant Growth Regulants", Ph.D. Dissertation, Iowa State Univ., Ames, Iowa (1952).
3. Hoffmann, O. L., Plant Physiol. 28, 622 (1953).
4. Woods, D. D., Brit. J. Expt. Path. 21, 74 (1940).
5. Hoffmann,O. L., P. W. Gull, H. C. Zeisig, and J. R. Epper-ly, Proc. North Central Weed Control Conf. 17,20 (1960).
6. Overbeek, J. van, in "The Physiology and Biochemistry of Herbicides" (Audus, L. J., Ed.), p. 387, Academic Press, New York, 1964.
7. Hoffmann, O. L., U. S. Patent 3,131,509 (1964).
8. Hoffmann,O. L., Weed Soc. of Amer. Abstracts 54 (1961).
9. Hoffmann,O. L., Weeds 10, 322 (1962).
10. Bendixen, L. E., Weed Sci. 23, 445 (1975).
11. Hoffmann,O. L., U. S. Patent 3,564,768 (1971).
12. Hoffmann,O. L., Ger. Offen. Patent 2,245,471 (1973).
13. Blair, A. M., C. Parker, and L. Kasasian, PANS 22, 65 (1976).
14. Mitchell, E. D., Jr., and N. E. Talbert, Biochemistry 7, 1019 (1968).
15. Ione, K., and R. Yamamote, Proc. Crop Sci. Soc. of Japan 44, 465 (1975).
16. Johnson, L. P. V., Can. J. Research (C) 13, 283 (1935).

17. Kahn, A. A., Physiol. Plant. 19, 869 (1966).
18. Fawcett, R. S., and F. W. Slife, Weed Sci. 23, 419 (1975).
19. Knudson, L., in "The Standard Cyclopedia of Horticulture" (Bailey, L. H., Ed.), p. 1146. Macmillan, New York, 1933.
20. Denny, F. E., Contrib. Boyce Thompson Inst. 1, 59 (1926).
21. Denny, F. E., Contrib. Boyce Thompson Inst. 1, 169 (1926).

Part II
Antidote Structure— Activity Relationships

ANTIDOTES PROTECT CORN FROM THIOCARBAMATE HERBICIDE INJURY[1]

Ferenc M. Pallos[2], Reed A. Gray[2], Duane R. Arneklev[2],
and Mervin E. Brokke[3]
Stauffer Chemical Co.

*N,N-Diallyl-2,2-dichloroacetamide and related compounds
added in small amounts to EPTC (S-ethyl dipropylthiocarbamate)
or other thiocarbamate herbicides prevent the onset of herbicide injury to corn plants and greatly increase crop yields.
Antidotes of this type provide a novel method to obtain
greater selectivity and new crop uses for the nonpersistent
thiocarbamate herbicides.*

I. INTRODUCTION

A new concept in weed control involves the use of
antidotes to protect crops from herbicide injury (1,2). We
have recently found a new class of herbicide antidotes (3)
which are superior to previously described compounds for
protecting corn from injury by EPTC (S-ethyl dipropylthiocarbamate) and other thiocarbamate herbicides.

EPTC, an established herbicide, has many desirable
characteristics including low toxicity to mammals and wild-
life. It also undergoes rapid degradation in the environment so it is appropriate for use in a crop rotation sequence. Yet, at the levels applied for weed control, it is
frequently phytotoxic to corn plants limiting its usefulness
on this major crop.

---

[1]This paper appeared with the same title and authors in J.
Agr. Food Chem. 23, 821 (1975). It is presented here with
permission from the American Chemical Society.

[2]Western Research Centers, Richmond, California 94804.

[3]Eastern Research Center, Dobbs Ferry, N. Y. 10522.

II. DISCOVERY OF R-25788 AS AN EPTC ANTIDOTE

Antidote screening tests were carried out in the greenhouse by incorporating EPTC into loamy sand soil at an
excessive rate of 6.7 kg/ha, so that corn planted in the soil
was severely injured. The soil was placed in 21 x 31 x 9 cm
metal flats to a depth of 7 cm. Hundreds of compounds were
synthesized and tested for antidotal activity by coating
the corn seeds with up to 0.5% of the compound by weight of
the seeds and planting ten seeds per flat 2 cm deep in the
soil containing EPTC. The coating was done by placing 50
mg of the compound in a glass vial with 10 g of corn seeds,
sealing, and shaking the vial. After growing in the greenhouse for 2 weeks, corn plants were evaluated for injury.
The crop injury was rated as follows: the number of plants
which showed leaf-rolling and stem-twisting injury symptoms
in the treatment were multiplied by 100 and divided by the
number of plants in the treatment. Table 1 gives the results
of these tests with 16 compounds chosen from several hundred
known active structures.

The most active EPTC antidotes are the N,N-disubstituted
dichloroacetamides. The monochloroacetamides are generally
less active than the dichloroacetamides. A variety of substituents on the nitrogen atom including alkyl, haloalkyl,
alkenyl, and heterocyclic groups impart various degrees of
protective activity. Usually compounds having two substituents on the nitrogen atom are more active than those
with only one substituent.

Further tests at lower rates showed that N,N-diallyl-
2,2-dichloroacetamide (R-25788) is the most active in the
group and well suited for practical application. When
applied at a rate of only 0.1% by weight of the corn seed,
it provided complete protection from EPTC at 6.7 kg/ha. Of
even greater interest was the discovery that a mixture of
EPTC and R-25788 applied to the soil before the seeds were
planted still gave complete protection of corn without
affecting the control of weeds.

In a loamy sand soil, 0.035 kg/ha of R-25788 was sufficient to protect corn from 3.4 kg/ha of EPTC, while two-
and fourfold larger doses of R-25788 were needed for complete
protection when the herbicide level was increased two- and
fourfold, respectively. Thus, a linear relationship exists
between the amount of herbicide applied and the amount of
antidote required. Only about 1 part of antidote is needed
per 100 parts of EPTC to protect corn plants in the greenhouse. When applied alone, R-25788 has no effect on corn

TABLE 1

Effect of Mono- and Dichloroacetamides as Seed Treatments at About 0.5% by Weight in Protecting Corn From Injury by EPTC Incorporated in the Soil at 6.7 kg/ha

Substituents on nitrogen		Corn injury, %
R	R_1	
Monochloroacetamides: $CH_2ClC(O)NRR^1$		
H	CH_2CH_2Br	10
H	$-\overset{\displaystyle C_2H_5}{\underset{\displaystyle C_2H_5}{C}}-CN$	10
H	$-CH_2-$ (tetrahydrofuranyl)	10
H	Methallyl	0
H	tert-Butyl	0
Allyl	Allyl	30
Dichloroacetamides: $CHCl_2C(O)NRR^1$		
H	Allyl	10
H	$-\overset{\displaystyle C_2H_5}{\underset{\displaystyle C_2H_5}{C}}-CN$	10
H	Methallyl	0
H	$-CH_2-$ (tetrahydrofuranyl)	0
H	$\overset{\displaystyle C_2H_5}{\underset{\displaystyle C_2H_5}{}}$ (cyclohexenyl)	0
C_2H_5	C_2H_5	0
n-C_3H_7	n-C_3H_7	0
Allyl	Allyl	0[a]
Allyl	(cyclohexenyl)	0
i-C_3H_7	(cyclohexenyl)	0
Check (no antidote, EPTC alone)		90

[a] R-25788.

even at 5.6 kg/ha.

While the antidote protects corn from the action of
EPTC, it affords no such protection to 24 different species
of weeds even at the ratio of 1.1 kg/ha of antidote to 2.2
kg/ha of EPTC. These weeds include johnsongrass (<u>Sorghum</u>
<u>halepense</u>), nutsedge (<u>Cyprus</u> spp.), and wild cane (<u>Sorghum</u>
<u>bicolor</u>), which are of great economic importance worldwide.

III.   R-25788 ALSO ANTIDOTES CORN INJURY FROM OTHER THIO-
       CARBAMATE HERBICIDES

     The results of tests carried out to determine if
R-25788 protects corn from other thiocarbamate herbicides
are reported in Table 2. The results show that two other
commercial thiocarbamates behave like EPTC but antidote
protection to the corn plant is incomplete with one herbicide of the same chemical class.

TABLE 2

*Effect of the Antidote R-25788 on Corn Injury
Caused by Different Thiocarbamate Herbicides*

Herbicide, kg/ha	Antidote, kg/ha	Corn in-jury, %
Vernolate,[a] 6.7	0	90
Vernolate, 6.7	0.14	0
Butylate,[b] 8.9	0	20
Butylate, 8.9	0.14	0
Cycloate,[c] 6.7	0	90
Cycloate, 6.7	0.14	50

[a]<u>S</u>-Propyl dipropylthiocarbamate. [b]<u>S</u>-Ethyl diisobutylthiocarbamate. [c]<u>S</u>-Ethyl cyclohexylethylthiocarbamate.

IV. EFFECTIVENESS OF R-25788 IN FIELD TRIALS

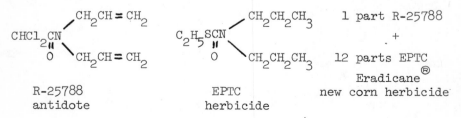

R-25788   EPTC   new corn herbicide
antidote   herbicide

The antidote N,N-diallyl-2,2-dichloroacetamide was introduced commercially in 1973 in the herbicide and antidote mixture Eradicane. The product contains 1 part of antidote for 12 parts of EPTC, clearly much more antidote than is needed in the greenhouse. This high margin of safety is employed to counter the unpredictable effects of the variable field environment. Data from Rains and Fletchall (4) show the remarkable effectiveness and crop safety of this antidote-EPTC mixture in Table 3. (For further information see also references 5-13).

TABLE 3

Effect of the Antidote on Corn Yields in Field Trials (4)

Treatment	Yield, kg/ha
Check (with weeds)	19.0
EPTC, 6.7 kg/ha	43.6
EPTC, 6.7 kg/ha, + R-25788, 0.56 kg/ha	98.6
Handweeded control	96.5

V.   PROPERTIES AND IMPORTANCE OF R-25788

R-25788 is relatively nontoxic to laboratory animals by acute oral or dermal application. The acute oral LD_{50} for rats is 2000 mg/kg and the acute dermal LD_{50} for rabbits is greater than 4640 mg/kg. It is not an eye irritant for rabbits. R-25788 undergoes rapid metabolism in soil, plants, and animals.

At the rates used, R-25788 is not phytotoxic; yet, when it is added in small amounts to a class of known herbicides and the mixture is applied in the field, this antidote alleviates the undesired phytotoxic effects of these herbicides. The antidote-EPTC mixture is very effective in weed control and safe on corn, the largest crop in the United States and second largest in the world.

VI.   REFERENCES

1.   Hoffman, O. L., Weeds 10, 322 (1962).
2.   Hoffman, O. L., Abstracts of the Meeting of the Weed
         Science Society of America, No. 12, 1969.
3.   Stauffer Chemical Co., Belgian Patent 782,120 (1972).
4.   Rains, L. J., and Fletchall, O. H., presented at the
         Meeting of the Weed Science Society of America,
         Atlanta, Ga., 1973.

5. Rains, L. J., and Fletchall, O. H., Proc. North Cent. Weed Control Conf. 26, 42 (1971).
6. Chang, F. Y., Bandeen, J. D., and Stephenson, G. R., Can. J. Plant Sci. 52, 707 (1972).
7. Chang, F. Y., Bandeen, J. D., and Stephenson, G. R., Weed Res. 13, 399 (1973).
8. Chang, F. Y., Stephenson, G. R., and Bandeen, J. D., Weed Sci. 21, 292 (1973).
9. Chang, F. Y., Stephenson, G. R., and Bandeen, J. D., J. Agr. Food Chem. 22, 245 (1974).
10. Heikes, P. E., and Swink, J. F., Proc. West. Soc. Weed Sci. 26, 32 (1973).
11. Schmer, D. A., Lee, G. A., and Alley, H. P., Proc. West. Soc. Weed Sci. 26, 38 (1973).
12. Smith, W. F., McAvoy, W. J., Lay, M. M., and Ilnicki, R. D., Proc. Northeast. Weed Sci. Soc. 27, 57 (1973).
13. Spotanski, R. F., and Burnside, O. C., Weed Sci. 21, 531 (1973).

Addendum. The patent in reference 3 is the same as:

Pallos, F. M., Brokke, M. E.,and Arneklev, D. R., U. S. Patent 4,021,224 (1977).

DIAZOSULFONATES AS PROTECTANTS AGAINST ATRAZINE TOXICITY TO SOYBEANS

B. A. Phillips and A. S. Bhagsari

Fort Valley State College
Fort Valley, Georgia

A series of benzenediazosulfonates were synthesized and evaluated as potential antidotes to residual atrazine toxicity to soybeans [Glycine max (L.) Merr.]. Greenhouse experiments were conducted using soil incorporated or seed coated sodium p-(dimethylamino)benzenediazosulfonate (dexon), sodium p-methylbenzenediazosulfonate (MBDS), sodium p-chlorobenzenediazosulfonate (CBDS), and sodium benzenediazosulfonate (BDS). A concentration of 0.2 ppm (w/w) of atrazine in the soil gave an average soybean dry weight of 16% of control after 26 days. Incorporation of dexon from 25 to 400 ppm (w/w) in soil containing 0.2 ppm atrazine gave proportionate increases in plant dry weight, being 76% of control at 400 ppm. After 66 days dexon completely prevented the reduction in soybean leaf and stem dry weight caused by atrazine, with similar protection for roots. Soil incorporated MBDS also protected soybeans, however, CBDS or BDS gave no protection against atrazine toxicity. Seed coating of the diazosulfonates did not provide protection, but a seed coating of 0.1% BDS alone gave a 27% increase in dry weight over control after 26 days.

I.   INTRODUCTION

The s-triazine herbicides (1) are widely used for pre-emergence and postemergence selective weed control in agriculture. Their principal mode of action is through the inhibition of the electron transport in photosystem II (2). Atrazine (Fig. 1) is a 2-chloro-s-triazine herbicide that is used extensively for weed control in corn. In the Midwest and Great Plains regions of the United States, and in Canada, residual amounts of atrazine can persist in the soil (3) and

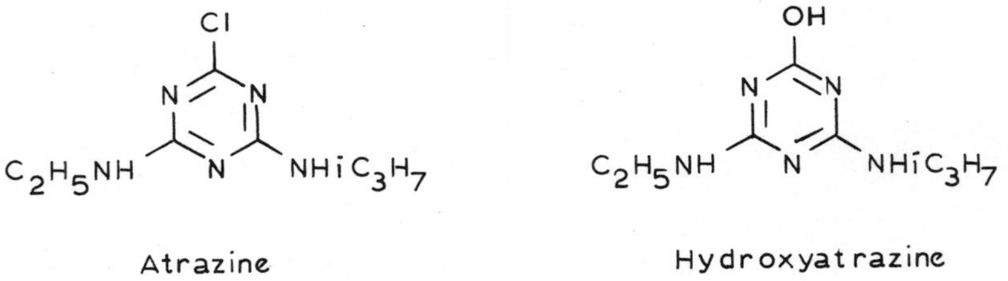

Atrazine   Hydroxyatrazine

Fig. 1.   Structure of atrazine [2-chloro-4-(ethylamino)- 6-(isopropylamino)-s-triazine] and hydroxyatrazine.

damage susceptible crops, such as sugar beets, small grains and soybeans grown in rotation with corn (4, 5, 6, 7).   This carryover effect limits the utility of atrazine in cropping sequences.

The toxic atrazine residue is partly due to the inability of the soil environment to chemically hydrolyze the herbicide to the nonphytotoxic hydroxyatrazine (Fig. 1) at a suitable rate (3).   Castelfranco and Brown (8) reported that pyridine and hydroxylamine catalyzed the hydrolysis of simazine (2-chloro-4,6-bis-ethylamino-s-triazine) in solution.   Sodium and calcium polysulfides were found to catalyze the detoxification of simazine in the soil (9), while the growth retardant CCC [(2-chloroethyl)trimethylammonium chloride] provided limited protection to wheat against subsequently applied atrazine (10). Behrens and coworkers (11) studied the effect of sublethal amounts of atrazine in combination with CDAA (2-chloro-N,N-diallylacetamide), linuron [3-(3,4-dichlorophenyl)-1-methyl-urea], amiben (3-amino-2,5-dichlorobenzoic acid), or trifluralin (α,α,α-trifluoro-2,6-dinitro-N,N-dipropyl-p-toluidine) on the growth of soybeans.   The effect of atrazine in combination with CDAA or linuron was not significant;   however, the growth of soybean seedlings was depressed by treatments containing combinations of trifluralin and atrazine.   The best growth was observed with combinations of atrazine and amiben.

An important mechanism by which corn detoxifies atrazine has been attributed to the presence of 2,4-dihydroxy-3-keto-7-methoxy-1,4-benzoxazine, which is originally present in the plant in the form of its 2-glucoside (Fig. 2;   R=glucose) (12, 13, 14, 15).   Tipton and coworkers (16) have studied the reaction of simazine with the benzoxazine derivative (Fig. 2; R=H) and suggested that a molecular aggregate of this hydroxamic acid may catalyze the hydrolytic detoxification reaction. Smissman *et al.*   (17) showed that N-hydroxysuccinimide and 1-hydroxy-2-piperidone (Fig. 2), which are model compounds of the natural corn resistance factor, nucleophilically catalyzed

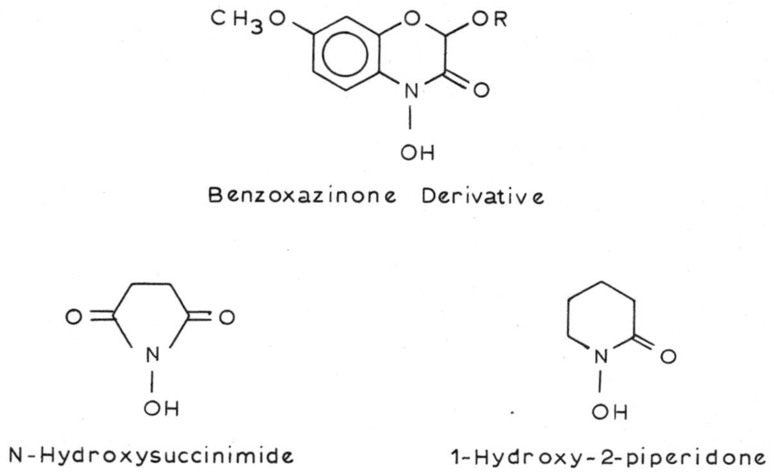

Benzoxazinone Derivative

N-Hydroxysuccinimide   1-Hydroxy-2-piperidone

Fig. 2. Structure of 2,4-dihydroxy-3-keto-7-methoxy-1,4-benzoxazine (R=H), N-hydroxysuccinimide, and 1-hydroxy-2-piperidone.

the hydrolysis of cyprazine (2-chloro-4-isopropylamino-6-cyclo-propylamino-s-triazine).

The 2-chloro-s-triazine herbicides will react with sulfonic acids in wet chloroform to form the sulfonate salt of the corresponding nonphytotoxic hydroxyatrazine (Fig. 3). However, under similar condition, no reaction was observed with the conjugate base (sodium salt) of the sulfonic acids (18). In our laboratory, a series of sulfonic acid derivatives (sodium benzenesulfonate, p-([(p-dimethylamino)phenyl]-azo)benzenesulfonic acid sodium salt, p-phenylazobenzenesulfonic acid, sodium p-acetylbenzenesulfonate, 1-naphthalenesulfonic acid, 1-naphthalenesulfonic acid sodium salt, and 1,5-naphthalenedisulfonic acid disodium salt) gave no protection to soybeans against atrazine toxicity. The fungicide dexon [sodium p-(dimethylamino)benzenediazosulfonate], which is used to control damping-off, has been reported to be antagonistic to triazine toxicity to cucumbers, oats and soybeans (19, 20, 21).

There has been very little success in finding antidotes for herbicides that inhibit photosynthesis, especially those, like atrazine, that are photosystem II (Hill reaction) inhibitors. This paper will present the initial results of the evaluation of a series of selected diazosulfonates (including

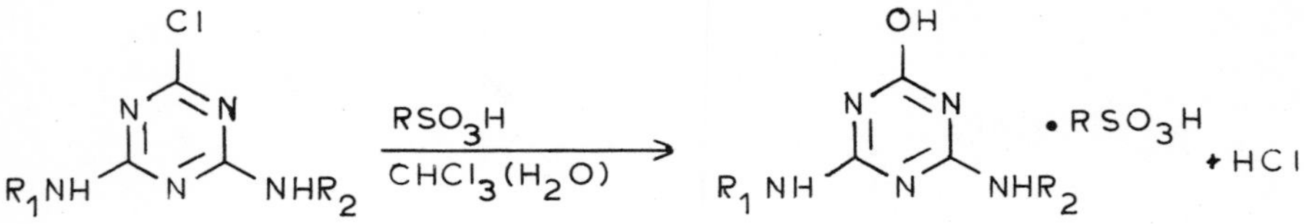

Atrazine   (R_1 = ethyl R_2 = isopropyl)
Cyprazine   (R_1 = isopropyl R_2 = cyclopropyl)
GS13529   (R_1 = ethyl R_2 = $\underline{t}$ - butyl)

Fig. 3.   *Reaction of 2-chloro-s-triazines with methane-sulfonic acid (R=CH₃) or benzenesulfonic acid (R=C₆H₅).*

dexon) as antidotes to residual atrazine toxicity to soybeans.

II. EXPERIMENTAL PROCEDURES

All experiments were conducted in pots in a greenhouse using soybeans (cultivar Bragg) and a Norfolk sandy loam soil (75.0% sand, 20.4% silt, 4.6% clay) with a pH of 6.2 and organic matter content of 1.15%. Atrazine was obtained in pure form by three crystallizations of commercial Aatrex 80 W with chloroform. Atrazine was dissolved in acetone and thoroughly mixed with one pound of soil. This soil was then incorporated into the bulk soil by using a soil mixer to give the desired concentration. Preliminary experiments indicated that a concentration of 0.2 ppm atrazine would give a soybean dry weight

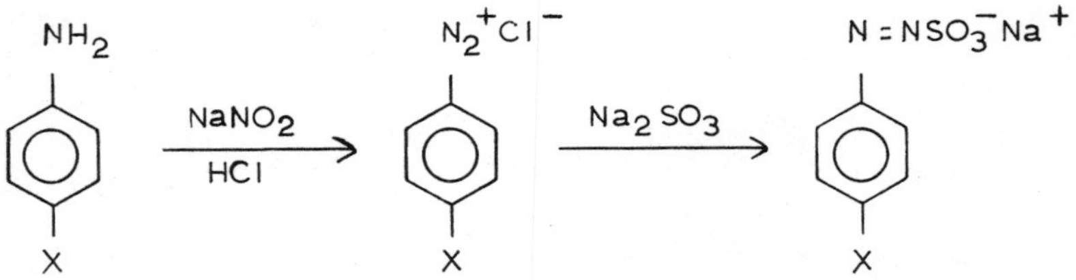

X = CH₃   MBDS
X = Cl   CBDS
X = H   BDS

Fig. 4.   *The synthesis of MBDS, CBDS, and BDS.*

(excluding roots) of 16% of control after 26 days of growth.
A concentration of 0.2 ppm atrazine was used in all experiments.

MBDS, CBDS and BDS were synthesized by the method of
Dijkstra and de Jonge (22). Dexon (Fig. 4; X=N(CH3)2) was
obtained from Chemagro. The concentration of all chemicals
was based on active ingredient. The diazosulfonates were
either soil incorporated or seed coated. Soil incorporation
was done by using a soil mixer and the seeds were coated by
shaking the seeds with the diazosulfonate in a flask.

III. SOIL INCORPORATION OF DIAZOSULFONATES

The initial approach was to incorporate the diazosulfonates into soil containing 0.2 ppm atrazine and determine the
effect on soybean growth. Dexon concentrations of 5, 10 and
15 ppm gave no protection against atrazine toxicity to soybeans. Subsequently, the concentration range was increased to
25-200 ppm. At the higher concentrations dexon alone caused
substantial decreases in plant dry weight after 26 days of
growth. The average dry weight with 100 and 200 ppm of soil
incorporated dexon was 64 and 60% of control plants, respectively (Fig. 5). This reduction in dry weight may have been

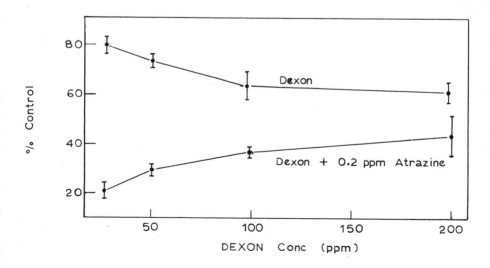

*Fig. 5. The effect of dexon alone and in the presence of
atrazine on soybean dry weight (excluding roots) after 26 days.*

TABLE 1

The Effect of Dexon and MBDS on Soybean
Plant Height after 26 Days

	Plant Height (inches)	
Concentration (ppm)	Dexon	MBDS
25	9.8 + 1.0	10.3 + 0.3
50	9.3 + 0.2	10.9 + 1.0
100	8.0 + 0.5	11.0 + 0.8
200	7.7 + 1.2	10.7 + 0.5
Control	10.7 + 0.2	10.6 + 0.8

due to a plant stunting effect caused by dexon, which became
more severe with increasing concentrations (Table 1). At 200
ppm dexon a 35% reduction in plant height was observed. The
addition of 25-200 ppm dexon to soil containing 0.2 ppm atrazine caused a proportionate increase in plant dry weight (Fig.
5). The yield was 20 and 43% of control for 25 and 200 ppm
dexon, respectively. Similar trends were observed for soybean
fresh weight (Table 2), leaf dry weight (Table 3), and stem
dry weight (Table 4). The leaf area data (Table 5) was consistently higher than that of the corresponding leaf dry
weight. This indicates that leaf expansion occurred, however,
in the presence of atrazine leaf photosynthate accumulation

TABLE 2

The Effect of Diazosulfonates Alone and in the Presence of
0.2 ppm Atrazine on Soybean Fresh Weight
(% of Control) After 26 Days

	Diazosulfonate		
Treatment	Dexon	MBDS	CBDS*
25 ppm	86.4 + 7.9	97.2 + 6.9	76.3 + 7.8
50 ppm	80.0 + 6.5	94.1 + 3.6	77.5 + 3.7
100 ppm	74.5 + 3.8	98.1 + 5.0	80.6 + 6.7
200 ppm	63.7 + 0.2	86.5 + 4.1	74.8 + 5.2
Atrazine alone	29.0 + 3.5	47.5 + 8.6	49.8 + 5.8
25 ppm + atrazine	39.6 + 5.3	48.8 + 9.3	37.5 + 8.5
50 ppm + atrazine	50.4 + 3.4	44.9 + 8.7	30.8 + 5.2
100 ppm + atrazine	56.1 + 2.7	53.5 + 7.6	49.4 + 10.2
200 ppm + atrazine	58.1 + 5.1	73.7 + 9.6	43.1 + 5.5

* Based on 32 days

TABLE 3

*The Effect of Diazosulfonates Alone and in the Presence of
0.2 ppm Atrazine on Soybean Leaf Dry Weight
(% of Control) After 26 Days*

| Treatment | Diazosulfonates | | |
	Dexon	MBDS	CBDS*
25 ppm	91.6 + 10.1	102.4 + 8.1	69.3 + 9.1
50 ppm	74.4 + 2.2	99.1 + 7.2	64.3 + 3.5
100 ppm	65.6 + 2.6	102.4 + 9.5	66.8 + 4.7
200 ppm	57.8 + 2.2	84.8 + 4.6	56.3 + 8.8
Atrazine alone	11.2 + 1.8	14.9 + 5.2	28.6 + 7.2
25 ppm + atrazine	20.2 + 3.7	27.8 + 8.4	20.4 + 8.3
50 ppm + atrazine	31.4 + 3.1	23.8 + 5.8	9.5 + 2.4
100 ppm + atrazine	40.0 + 0.6	30.9 + 7.1	30.5 + 8.6
200 ppm + atrazine	46.2 + 3.5	50.1 + 9.2	29.5 + 4.1

*Based on 32 days

TABLE 4

*The Effect of Diazosulfonates Alone and in the Presence of
0.2 ppm Atrazine on Soybean Stem Dry Weight
(% of Control) After 26 Days*

| Treatment | Diazosulfonate | | |
	Dexon	MBDS	CBDS*
25 ppm	78.7 + 4.4	92.7 + 7.9	73.4 + 8.8
50 ppm	70.1 + 5.2	90.9 + 6.4	68.6 + 7.8
100 ppm	62.6 + 8.3	85.6 + 7.1	77.4 + 7.4
200 ppm	63.4 + 9.4	78.8 + 5.6	68.6 + 9.5
Atrazine alone	16.3 + 1.6	21.1 + 7.3	27.5 + 5.2
25 ppm + atrazine	21.0 + 3.2	28.5 + 6.0	17.8 + 5.2
50 ppm + atrazine	29.9 + 2.1	26.3 + 5.5	15.6 + 3.1
100 ppm + atrazine	36.2 + 3.1	31.3 + 4.7	33.3 + 9.3
200 ppm + atrazine	40.1 + 5.9	43.9 + 9.2	28.4 + 5.1

*Based on 32 days

lagged behind.

In order to determine the amount of dexon necessary to prevent atrazine damage to soybeans, the dexon concentration was increased to 400 ppm and the plants were grown for 66 days (Fig. 6). The reduction in dry weight with increasing levels

TABLE 5

The Effect of Diazosulfonates Alone and in the Presence of
0.2 ppm Atrazine on Soybean Leaf Area
(% of Control) After 26 Days

Treatment	Diazosulfonate		
	Dexon	MBDS	CBDS*
25 ppm	89.5 + 3.9	92.1 + 9.4	74.8 + 9.6
50 ppm	85.8 + 7.5	95.7 + 6.5	70.1 + 3.4
100 ppm	73.4 + 8.1	95.1 + 2.8	71.8 + 4.9
200 ppm	68.5 + 4.4	85.7 + 4.3	56.6 + 8.1
Atrazine alone	21.7 + 4.2	27.7 + 8.5	48.2 + 8.7
25 ppm + atrazine	38.0 + 6.5	38.7 + 11.4	18.4 + 6.5
50 ppm + atrazine	54.4 + 3.5	48.4 + 4.1	14.1 + 8.2
100 ppm + atrazine	67.0 + 0.2	58.1 + 2.1	34.2 + 2.8
200 ppm + atrazine	69.7 + 6.5	79.3 + 8.5	35.8 + 9.8

*Based on 32 days.

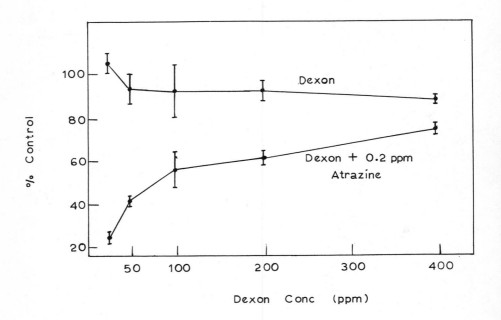

Fig. 6. The effect of dexon alone and in the presence of
atrazine on soybean dry weight (excluding roots) after 66 days.

of dexon alone was less for 66-day old plants than 26-day old plants. The average dry weight of plants grown for 66 days with 400 ppm dexon was 84% of control with dexon alone and 76% with both atrazine and dexon present. At higher concentrations of dexon the reduction in dry weight of soybeans from atrazine toxicity was essentially prevented. The typical symptoms of atrazine injury (leaf chlorosis followed by necrosis) were observed with decreasing severity up to 200 ppm of dexon, however, no signs of atrazine toxicity were evident at 400 ppm dexon. This indicated either less atrazine uptake in the presence of dexon, a concept that has been previously reported (23), or its detoxification in soil or the plant system. We reported (18) that atrazine does not react with dexon in acetonitrile or water solutions in the flask, however, this does not preclude a soil catalyzed reaction.

Soil incorporation of dexon alone up to 400 ppm caused no reduction in total root dry weight relative to control (Fig. 7). At 400 ppm dexon completely protected the roots from atrazine damage in terms of dry weight. Also, at the highest dexon concentration the dry weight of the total roots (95% of control) was substantially higher than that of the corresponding aerial parts (76%), indicating a possible preferential flow of photosynthate to the roots. No distinction was made between primary and secondary roots in this study.

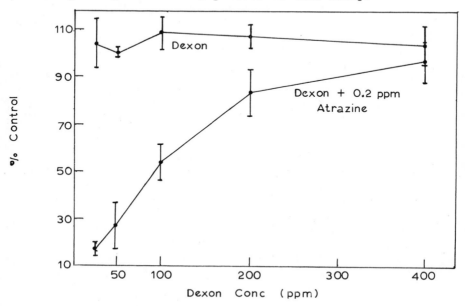

Fig. 7. The effect of dexon alone and in the presence of atrazine on soybean total root dry weight after 66 days.

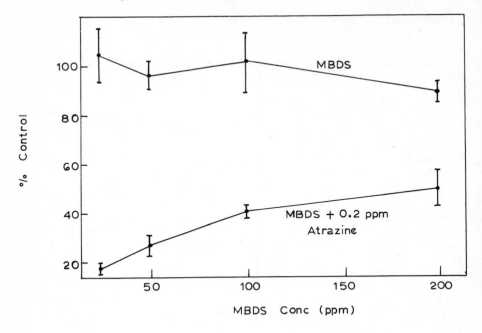

Fig. 8. The effect of MBDS alone and in the presence of atrazine on soybean dry weight (excluding roots) after 26 days.

Application of MBDS alone (Fig. 8) showed less depressing effect upon soybean growth than dexon (Fig. 5) after 26 days. No reduction in dry weight was observed up to 100 ppm MBDS, but at 200 ppm the dry weight was 82% of control plants. MBDS had no effect upon plant height (Table 1). Like dexon, 25-200 ppm MBDS in the presence of atrazine gave a gradual increase in plant dry weight, with a maximum of 52% of control at 200 ppm. Symptoms of atrazine toxicity were observed in all plants grown in MBDS-treated soil, however, the extent of injury was much less at higher concentrations. The antidotal effects of soil incorporated MBDS on atrazine toxicity to soybeans were similar in trend to dexon for other growth parameters studied and are summarized in Tables 2-5.

Soil incorporated CBDS and BDS alone and in combination with atrazine caused a drastic reduction in soybean dry weight (Table 6). The plants were severely stunted and were characterized by shrivelling of the leaves. BDS alone induced branching at the unifoliate leaf nodes, indicating an effect on plant growth regulating processes. No protection against atrazine toxicity to soybeans was observed with soil incor-

TABLE 6

The Effect of BDS and CBDS Alone and in the Presence of
0.2 ppm Atrazine on Soybean Dry Weight (% of Control)
After 26 and 32 Days, Respectively

Treatment	Diazosulfonates	
	CBDS	BDS
25 ppm	67.2 + 7.4	79.5 + 4.7
50 ppm	66.2 + 4.2	62.7 + 7.3
100 ppm	71.4 + 5.8	55.8 + 7.6
200 ppm	65.8 + 4.8	40.3 + 5.6
Atrazine alone	25.6 + 3.0	11.9 + 4.9
25 ppm + atrazine	15.9 + 4.9	7.9 + 2.2
50 ppm + atrazine	11.2 + 3.1	6.9 + 1.2
100 ppm + atrazine	21.5 + 5.5	9.8 + 2.0
200 ppm + atrazine	28.5 + 5.2	8.9 + 2.7

porated CBDS (Tables 2-6) or BDS (Table 6).

The protection provided to soybeans by the diazosulfonates is summarized in Figure 9. Protection here means the
"percent effect" or the dry matter yield (% of control) from
the antidote + atrazine treatment less the dry matter from the
atrazine alone treatment. Dexon consistently gave higher protection than all other diazosulfonates at all levels. At 400
ppm the protection provided by dexon was 43% after 66 days.
MBDS gave a slight protection at 200 ppm, however, only about
half of that provided by dexon. BDS gave no protection, while
CBDS at low concentrations seemed to be synergistic with the
injurious effects of atrazine. The mode of action of the
antidotal activity of dexon and MBDS is not completely understood. The relatively large amounts of diazosulfonates required to counteract the toxicity of 0.2 ppm atrazine indicates a low probability of a detoxification *via* a chemical
reaction in the soil. Further, the addition of 25-400 ppm
dexon did not alter soil pH after 66 days.

IV.   SEED COATING OF DIAZOSULFONATES

The concentrations of the chemicals applied in seed coating experiments ranged from 0.05 to 0.5% (seed weight basis)
and were varied according to their effect upon germination.
Seed coating above 0.1% dexon caused a drastic reduction
in percent germination, being 39% at 0.5% dexon. The diazosulfonates alone had no inhibitory effect upon plant growth
(Table 7). In one experiment 0.1% BDS alone caused a 27% increase in dry matter yield after 26 days. Seed coated dexon,

TABLE 7

The Effect of Seed Coated Diazosulfonates Alone and in the Presence of 0.2 ppm Atrazine on Soybean Dry Weight (% of Control) After 26 Days

Treatment	Diazosulfonate			
	Dexon	MBDS	CBDS	BDS
0.05%	93.2 ± 5.2	107.1 ± 8.3	----	----
0.1%	102.7 ± 11.8	113.0 ± 7.7	101.6 ± 6.4	127.2 ± 5.9
0.15%	----	106.2 ± 7.1	----	----
0.3%	----	----	102.8 ± 9.3	110.3 ± 3.3
0.5%	----	----	101.1 ± 9.7	107.0 ± 3.7
0.2 ppm atrazine	27.4 ± 4.8	21.9 ± 9.1	20.6 ± 4.9	13.2 ± 1.0
0.05% + 0.2 ppm atrazine	28.4 ± 3.9	10.0 ± 1.0	----	----
0.1% + 0.2 ppm atrazine	41.9 ± 6.8	15.8 ± 2.8	23.4 ± 9.6	16.1 ± 1.8
0.15% + 0.2 ppm atrazine	----	9.1 ± 2.3	----	----
0.3% + 0.2 ppm atrazine	----	----	15.5 ± 2.6	14.3 ± 3.5
0.5% + 0.2 ppm atrazine	----	----	25.2 ± 7.7	38.3 ± 3.8

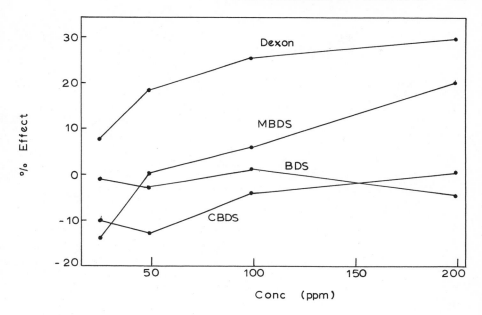

Fig. 9. Protective effects of diazosulfonates against atrazine toxicity to soybeans after 26 days (32 days for CBDS).

MBDS, and CBDS gave no substantial protection to soybeans in the presence of atrazine, however, 0.5% BDS showed a 22.2% increase in dry weight over atrazine alone. The effect of 0.1% BDS as a seed coating (Table 7) on soybean dry weight is the reverse of that observed in soil incorporation experiments (Table 6). This indicates that at low concentrations BDS has a growth promoting effect, while at high concentrations it acts as an inhibitor. The growth regulating aspects of BDS are presently under further investigation.

V.   REFERENCES

1.   For a review see, "Residue Reviews, The Triazine Herbicides" (F. A. Gunther, and J. D. Gunther, Eds.), Vol. 32, Springer-Verlag, New York, N. Y., 1970.
2.   Shimabukuro, R. H., and Swanson, H. R., J. Agr. Food Chem. 17, 199 (1969).
3.   Sheets, T. J., in "Residue Reviews, The Triazine Herbicides", Vol. 32, p. 287. Springer-Verlag, New York, N. Y., 1970.

4. Saghir, A. R., and Choudhary, A. H., Weed Res. 7, 272 (1967).

5. Frank, R., Weeds 14, 82 (1966).

6. Fink, R. J., and Fletchall, O. H., Weed Sci. 17, 35 (1969).

7. Buchholtz, K. P., Weeds 13, 362 (1965).

8. Castelfranco, P., and Brown, W. S., Weeds 10, 131 (1962).

9. Castelfranco, P., and Deutsch, D. B., Weeds 10, 244 (1962).

10. Kirkland, K., Diss. Abstr. B 33, 4066 (1973).

11. Espinoza, W. G., Adams, R. S., and Behrens, R., Agron. J. 60, 183 (1968).

12. Virtanen, A. I., and Hietala, P. K., Syom. Kemistilehti B 32, 252 (1959).

13. Wahlroos, O., and Virtanen, A. I., Acta Chem. Scand. 13, 1609 (1959).

14. Hietala, P. K., and Virtanen, A. I., Acta Chem. Scand. 14, 502 (1960).

15. Honkanen, E., and Virtanen, A. I., Acta Chem. Scand. 14, 504 (1960).

16. Tipton, C. L., Husted, R., and Tsao, F., J. Agr. Food Chem. 19, 484 (1971).

17. Nakano, N., Smissman, E. E., and Schowen, R., J. Org. Chem. 38, 4396 (1973).

18. Phillips, B. A., Juenge, E. C., and Brown, C. B., Ft. Valley St. Coll. Res. Bull. 1, 24 (1976).

19. Corbin, F. T., and Sheets, T. J., Weed Sci. Soc. Amer. Abstr., p. 22, 1968.

20. Nash, R. G., and Harris, W. G., Weed Sci. Soc. Amer. Abstr., P. 240, 1969.

21. Webster, H. L., and Sheets, T. J., Weed Sci. Soc. Amer. Abstr., p. 239, 1969.

22. Dijkstra, R., and de Jonge, J., Rec. trav. chim. 77, 538 (1958).

23. Webster, H. L., Dissertation "The Response of Plants to Atrazine-Dexon Combinations and the Influence of Dexon on the Absorption, Distribution and Metabolism of Atrazine by Plants", University Microfilms, Inc., Ann Arbor Michigan, 1969.

Part III

Comparative Activity, Selectivity, and Field Applications of Herbicide Antidotes

COMPARATIVE ACTIVITY AND SELECTIVITY OF HERBICIDE ANTIDOTES

G.R. Stephenson and F.Y. Chang[1]
Department of Environmental Biology
University of Guelph

*N,N-Diallyl-2,2-dichloroacetamide (R-25788) is physiolo-
gically selective as an antidote for S-ethyl dipropylthiocarbamate (EPTC) in corn (Zea mays L.) but not in other plant
species. 1,8-Naphthalic anhydride (NA) is much less active
in soil than R-25788 as an antidote for EPTC in corn but
highly active for preventing EPTC injury in corn and other
plant species if applied as a seed treatment. R-25788 was
shown to be highly active as an antidote for several other
herbicides in addition to EPTC in corn. NA also counteracted
the effects of several different herbicides in corn and other
plant species. One of the more useful applications of NA was
as a seed treatment to prevent barban (4-chloro-2-butynyl m-
chlorocarbanilate) injury to oats (Avena sativa L.), a combination that could permit the selective chemical control of
wild oats (Avena fatua L.) in oats. R-25788 clearly does not
act by preventing uptake, translocation or metabolism of EPTC
in corn seedlings. The high activity of several analogues of
R-25788 as antidotes for EPTC which were chemically very similar to the herbicide should have importance with respect to
the mode of antidote action and could indicate a possible
approach to finding antidotes for other herbicides in plants.*

I. INTRODUCTION

Since the discovery of herbicidal activity for nitro-
phenols in 1935 (1) and phenoxyalkanoic acids during World
War II (2, 3), over 100 different organic chemicals have been
developed for the selective control of many problem weeds in
most of the world's major crops. There have been two basic
approaches to selective chemical weed control (4). Firstly,
physical barriers have been employed to allow contact of the
weeds but not the crops with rather non-selective herbicides.

---

[1]Present address: Agriculture Canada, Pest Control Products
Division, Ottawa, Canada.

Secondly, many herbicides have been developed which are highly selective because morphological or physiological differences result in toxic amounts of the herbicide reaching the
site(s) of action in many weed species but not in certain
crops (4). Highly selective herbicides such as phenoxy herbicides in cereal crops, nitroanilines and thiocarbamates in
many broadleaved crops, and triazines in corn have resulted
in remarkable gains in crop yields. However, we have now
experienced several years of repeated use with many selective
herbicides and in many crops it is obvious that our chemical
methods are far less than perfect. In some cases the crop is
only marginally tolerant to the herbicides required for the
major weed problems and crop injury can often occur due to
unusual weather conditions, interactions with other chemicals,
crop varietal differences or errors in application. In other
situations, the crop may be highly tolerant to major herbicides but repeated use of these herbicides results in weed
population shifts to weed species that are as tolerant to
these herbicides as the crop itself. Thus the most difficult
to control weeds in cereals are 2,4-D resistant annual grasses, particularly wild oats; in corn, atrazine resistant
annual grasses and sedges; in winter rapeseed (*Brassica napus*
L.), wild mustards; and in tomatoes (*Lycopersicon esculentum*
Mill.), wild nightshades. While the chemical industry will
continue its search for new herbicides, the discovery of compounds with the high degree of selectivity required seems un-
likely.

It is clear that a new approach to selective chemical
weed control is now needed. One approach that appears to be
particularly promising is the development of chemical methods
to selectively protect the crop from herbicides that would
normally cause crop injury. With such an approach, crop
tolerance in marginally selective situations can be improved
and physiologically similar weeds can be controlled. In 1962,
Hoffmann (5) introduced the idea of herbicide antidotes for
crops with his discovery of various chemicals that, applied
as seed treatments, reduced injury to wheat (*Triticum aestivum* L.) from later foliar applications of barban (4-chloro-2-
butynyl *m*-chlorocarbanilate). In 1969, Hoffman (6) reported
on another chemical antidote, NA (1,8-naphthalic anhydride)
which prevented injury to corn (*Zea mays* L.) from EPTC (*S*-
ethyl dipropylthiocarbamate). NA subsequently became the
first chemical antidote to be used commercially to improve
herbicide selectivity in crops. Then in 1972, R-25788 (N,N-
diallyl-2,2-dichloroacetamide) was developed by Pallos,
Brokke, and Arneklev (7) as an antidote for EPTC in corn that
could be applied to the soil with the herbicide. Commercial

formulations of EPTC and butylate (S-ethyl diisobutylthiocarbamate) now contain R-25788 and can be applied at required rates for the control of difficult weeds with less risk of injury to corn.

II. PHYSIOLOGICAL SELECTIVITY OF N,N-DIALLYL-2,2-DICHLORO-ACETAMIDE (R-25788) AS AN ANTIDOTE FOR EPTC IN CORN

The antidotes developed by Hoffman for barban in wheat (5) and for EPTC in corn (6) were in all cases selectively applied to the crops as seed treatments. In reports of field studies during 1971, Bandeen (8) and Brown and Shaw (9) indicated that R-25788 was a very different kind of antidote in two respects. Firstly, as a preplant, soil incorporated application with EPTC, it effectively prevented EPTC injury to corn. Secondly, and even more surprising, was that as a broadcast application to the soil with EPTC, it did not reduce EPTC effectiveness for weed control. Thus R-25788 selectively protected the corn crop from EPTC without being selectively applied. This selective effect could have been related to soil factors such as differential movement of EPTC or R-25788 that could have resulted in corn seed at one depth being exposed to both herbicide and antidote and some weed seeds at a different depth being exposed to only the herbicide. To determine the role of soil factors in the selectivity of R-25788 as an antidote, both corn and green foxtail [*Setaria viridis* (L.) Beauv.] were grown in quartz sand nutrient culture (10). In this type of system, both plants could be equally exposed to toxic levels of EPTC with and without R-25788 in the nutrient solution.

In this nutrient culture system (Table 1), R-25788 provided nearly complete prevention of EPTC injury to corn with no significant effect for green foxtail. It can be concluded then, that the selectivity of R-25788 is physiological and not related to soil factors or to any type of direct interaction between the two chemicals occurring outside of the plant.

In other growth room studies (10) conducted with soil pretreated with toxic levels of EPTC, R-25788 (0.56 or 1.12 kg/ha) was examined as an antidote for EPTC injury in several other plant species. To obtain significant inhibition of shoot growth in all plant species, rates of EPTC application varied from 0.14 to 13.44 kg/ha. Twelve different crop and weed species were included in the study. R-25788 was completely inactive as an antidote for EPTC injury in all species except corn and sorghum (*Sorghum vulgare* Pers.) (Table 2).

TABLE 1

Differential Activity of R-25788 in Quartz Sand Nutrient Culture as an Antidote for EPTC Injury in Corn or Green Foxtail

Corn			Green Foxtail		
EPTC (ppm)	R-25788 (ppm)	Shoot Wt* (g)	EPTC (ppm)	R-25788 (ppm)	Shoot Wt* (g)
0	0	6.8 a	0	0	2.5 b
	1	7.2 a		1	3.0 a
	2	7.2 a		2	2.7 ab
10	0	3.1 d	0.4	0	0.5 c
	1	7.1 a		1	1.1 c
	2	7.5 a		2	1.2 c
20	0	2.1 c	1.0	0	0 d
	1	5.8 b		1	0 d
	2	5.9 b		2	0 d

* Data are the means of four replicates. Within the same column, means followed by the same letter do not significantly differ (5% level, Duncan's multiple range).

TABLE 2

Evaluation of R-25788 as an Antidote for EPTC in Various Plant Species in Addition to Corn

Species	Antidote Activity*
Corn (*Zea mays* L.)	+++
Sorghum (*Sorghum vulgare* Pers.)	+
Barley (*Hordeum vulgare* L.)	−
Rapeseed (*Brassica napus* L.)	−
Turnip (*Brassica rapa* L.)	−
White bean (*Phaseolus vulgaris* L.)	−
Crabgrass [*Digitaria sanguinalis* (L.) Scop.]	−
Green foxtail [*Setaria viridis* (L.) Beauv.]	−
Quackgrass [*Agropyron repens* (L.) Beauv.]	−
Yellow nutsedge (*Cyperus esculentus* L.)	−
Lambsquarters (*Chenopodium album* L.)	−
Redroot pigweed (*Amaranthus retroflexus* L.)	−

* +++, +, or − refer to complete, slight, and no reduction in EPTC injury to plant shoots with R-25788 at 0.56 or 1.12 kg/ha in combination with rates of EPTC which reduced shoot growth by more than 50% without R-25788.

These results further confirm that R-25788 is botanically specific as an antidote for EPTC and that its action may
be related to some physiological process in corn that may
occur to some lesser degree in the botanically similar sorghum plant.

III. COMPARATIVE SELECTIVITY OF NA (1,8-NAPHTHALIC ANHYDRIDE)
      AND R-25788 (N,N-DIALLYL-2,2-DICHLOROACETAMIDE) AS ANTI-
      DOTES FOR EPTC IN PLANTS

      In growth room studies (11) conducted in soil and quartz
sand nutrient culture, R-25788 and NA were compared as antidotes for corn and green foxtail to determine if NA was also
physiologically specific for corn as an antidote for EPTC.
As corn seed treatments (0.5% w/w) in soil (Fig. 1), NA and
R-25788 were equally effective as antidotes for low rates of
EPTC. However, at higher rates of EPTC (6.7-10.1 kg/ha), NA
was much less effective than R-25788 for reducing EPTC injury
to corn.

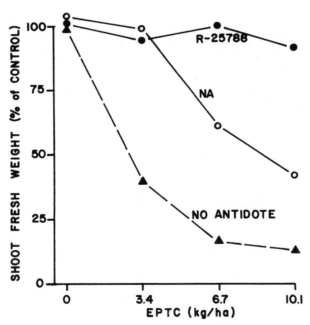

Fig. 1.   Influence of R-25788 or naphthalic anhydride
seed treatments (0.5% w/w) on EPTC toxicity to corn.

As broadcast applications to soil, an extremely high
rate of 28 kg/ha was required to reduce EPTC injury to corn
with NA and even then it was not as effective as was R-25788
at 0.6 kg/ha (Fig. 2). It is particularly significant that
when R-25788 or NA was applied to soil at rates sufficient to
reduce EPTC injury to corn, NA at these same rates also reduced injury to green foxtail. Similar results were obtained
in quartz sand nutrient culture (Fig. 3). With R-25788, no
significant reduction in EPTC injury to green foxtail was
observed in either soil or nutrient culture.

It is clear from these studies that NA and R-25788 have
different mechanisms as antidotes for EPTC in plants since
R-25788 is physiologically specific for protecting only corn
from EPTC while NA is not. Even if NA was more active in
soil, it would still be necessary to apply it selectively as
a seed treatment; otherwise, reduced weed control with EPTC
would result.

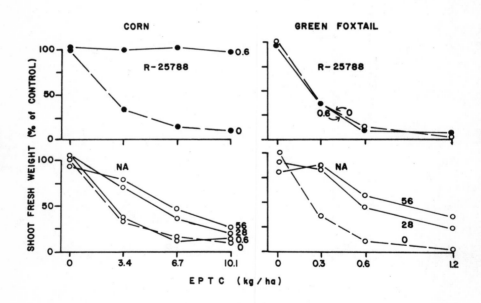

Fig. 2. Comparative activity of R-25788 and naphthalic
anhydride in soil as antidotes for EPTC toxicity in corn or
green foxtail. Rates of application were 0 and 0.6 kg/ha for
R-25788 and 0, 0.6, 28, and 56 kg/ha for NA.

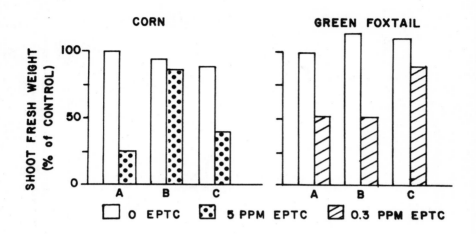

Fig. 3. Influence of two antidotes on EPTC toxicity to corn and green foxtail in quartz sand nutrient culture. (A) No antidote; (B) 2 ppmw of R-25788, (C) 2 ppmw of naphthalic anhydride. Both antidotes significantly reduced EPTC injury to corn (5% level); only naphthalic anhydride (C) significantly reduced EPTC injury to green foxtail.

IV. ACTIVITY OF R-25788 (N,N-DIALLYL-2,2-DICHLOROACETAMIDE) AND NA (1,8-NAPHTHALIC ANHYDRIDE) AS ANTIDOTES FOR OTHER HERBICIDES IN ADDITION TO EPTC

The antidote action of R-25788 in field corn (cv. United Hybrid 106) was examined for twenty-two herbicides with bio-assays conducted in growth rooms in a sandy loam soil (Table 3). R-25788 at 0 or 0.6 kg/ha was incorporated into the top 5 cm of soil in 10 cm (diameter) plastic pots. Preemergence herbicide treatments were applied as sprays to the soil surface immediately after planting the corn. Preplant soil incorporated herbicide treatments were applied as tank mixed sprays and incorporated with R-25788. Postemergence herbicide treatments were applied to the corn foliage 6-8 days after seeding when the corn seedlings were in the 2-3 leaf stage (12). Each herbicide was applied at 1, 2, and 3 times a basic rate of application.

From the results in this study (Table 3), it is obvious that while the expression of antidote activity for R-25788 may be botanically specific to corn, within corn it is not chemically specific as an antidote for only EPTC. In addition to being significantly active as an antidote for all

TABLE 3

Evaluation of R-25788 as an Antidote for 22 Different Herbicides in Corn

Herbicide	Chemical Name	Basic[1] Rate (kg/ha)	Application[2]	Antidote Activity of R-25788 % Reduction in Herbicide Injury[3]
Carbamate				
Barban	4-chlorobut-2-ynyl N-(3-chlorophenyl)carbamate	0.8	post	84
Chlorpropham	Isopropyl N-(3-chlorophenyl)carbamate	1.1	pre	-
Thiocarbamate				
Butylate	S-ethyl NN-di-isobutyl(thiocarbamate)	7.8	ppi	56
Di-allate	S-2,3-dichloroallyl NN-di-isopropyl(thiocarbamate)	5.6	ppi	26
EPTC	S-ethyl NN-dipropyl(thiocarbamate)	1.7	ppi	96
Molinate	S-ethyl NN-hexamethylene(thiocarbamate)	5.6	ppi	60
Pebulate	S-propyl N-butyl-N-ethyl(thiocarbamate)	5.6	ppi	40
Vernolate	S-propyl NN-dipropyl(thiocarbamate)	2.8	ppi	62
Dithiocarbamate				
Sulfallate	2-chloroallyl NN-diethyldithiocarbamate	5.6	ppi	68
Amide				
Alachlor	2,6-diethyl-N-(methoxymethyl)-α-chloroacetanilide	4.5	pre	50
Diphenamid	NN-dimethyl-αα-diphenylacetamide	2.5	pre	-
Pronamide	3,5-dichloro-N-(1,1-dimethylpropynyl)benzamide	0.3	pre	-
Propachlor	α-chloro-N-isopropylacetanilide	22.4	ppi	-
Others				
Aminotriazole	3-amino-1,2,4-triazole	2.2	post	-
Bromoxynil	3,5-dibromo-4-hydroxybenzonitrile	2.2	post	-
2,4-D	2,4-dichlorophenoxyacetic acid	4.5	pre	-
Dalapon	2,2-dichloropropionic acid	9.0	ppi	-
Dicamba	3,6-dichloro-2-methoxybenzoic acid	1.1	pre	-
Linuron	N'-(3,4-dichlorophenyl)-N-methoxy-N-methylurea	2.2	pre	32
MH	1,2,3,6-tetrahydro-3,6-dioxopyridazine	2.2	pre	-
Oryzalin	3,5-dinitro-N⁴N⁴-dipropylsulphanilamide	0.1	ppi	-
Trifluralin	2,6-dinitro-NN-dipropyl-4-trifluoromethylaniline	0.5	ppi	-

[1]Herbicides were applied at 1, 2, and 3 times this basic rate.
[2]Methods of application: ppi = pre-plant incorporated; pre = pre-emergence spray; post = post-emergence foliar spray.
[3]Per cent reduction in injury to corn shoots (fresh weight) for rates of the respective herbicides which reduced corn shoot growth to approximately 50% of the control. (-) refers to no reduction in herbicide injury with R-25788.

thiocarbamate herbicides in the study, R-25788 was also highly active for reducing injury to corn from sulfallate, a
dithiocarbamate; barban, a carbamate; and alachlor, an amide.
Barban was the only herbicide applied to corn foliage that
was effectively antidoted by a preplant, soil incorporated
application of R-25788.

Soil applications of R-25788 were found to be ineffective for reducing barban injury to oats (*Avena sativa* L.)
(Table 4). However, in separate studies (13), seed treatments with NA (0.5-1.0%) were highly effective for reducing
injury from foliar applications of barban (Fig. 4) to oats
and were significantly active for reducing diallate [S-(2,3-
dichloroallyl)diisopropylthiocarbamate] and triallate [S-
(2,3,3-trichloroallyl)diisopropylthiocarbamate] injury as
well (Table 5). Using NA as a seed applied antidote and barban applied postemergence, wild oats have now been selectively controlled in oats under actual field conditions. However,
factors responsible for variability in oat tolerance have to
be fully understood before the commercial feasibility of this
treatment can be determined.

TABLE 4

The Effects of R-25788 on the Toxicity of Barban to Oats
(cv. Garry)*

Barban (kg/ha)	Shoot Fresh Weight (% of Control) R-25788 (kg/ha)	
	0	0.6
0	100	97
0.42	61	67
0.84	19	16

* R-25788 was applied, preplant soil incorporated and barban
was applied postemergence during the two leaf stage. Means
within the same herbicide level (R-25788) did not significantly differ (0.05 level).

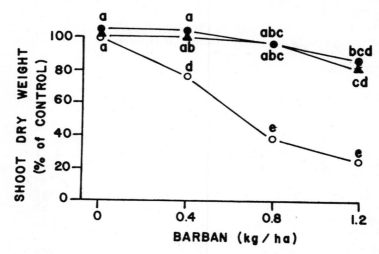

Fig. 4. Effect of seed treatment with NA on oat tolerance to barban. Means followed by the same letter do not
significantly differ (.05 level). O No NA; △0.5% NA; ●1.0% NA.

TABLE 5

Reduction of Diallate and Triallate Injury to Oats with NA
as a Seed Treatment

| | | Seed Treatment | |
| | | None | NA 1% by Seed Wt |
Herbicide	(kg/ha)	(0-10)*	(0-10)
Diallate	0	0 a	0 a
	0.35	7.1 d	2.4 b
	0.70	8.9 e	5.6 c
	1.40	9.9 f	9.0 e
Triallate	0	0 a	0 a
	0.35	7.0 d	3.0 b
	0.70	8.6 e	5.9 c
	1.40	9.7 f	8.6 e

* Observations were made 14 days after seeding. Injury
rating: 0 = none and 10 = kill. Means within the same herbicide group followed by the same letter are not significantly different (5% level, Duncan's multiple range).

In an excellent review article on herbicide protectants and antidotes, Blair *et al*. (14) have reported on a large number of herbicide-crop situations in which NA or R-25788 have significantly improved crop tolerance. In their summary of antidote research (Table 6), Blair *et al*. confirm the botanical specificity of R-25788, since it was not found to have good activity as an antidote for any crop except corn (maize). The non-selectivity of NA, as a seed applied antidote, was also confirmed since it was shown to have good activity for preventing herbicide injury to rice (*Oryza sativa* L.) and sorghum in addition to oats and corn (Table 6).

TABLE 6

Herbicides Counteracted by Antidotes and the Crops for which Each Antidote Provides Protection (14)

Herbicide	NA (As a Seed Dressing)		R-25788 (Sprayed onto Soil)	
	Good	Moderate	Good	Moderate
Alachlor	Sorghum Rice		Maize	
Barban	Oats Maize		Maize	Wheat
Butachlor		Rice		
Butylate		Maize		
CGA 24705		Rice		
Cisanilide		Maize Cotton Sorghum		
Di-allate	Sorghum	Oat		Maize
Dowco 221		Rice		
Epronaz		Rice		
EPTC	Maize	Maize Field bean	Maize	Maize Barley Field bean
Ethofumesate		Rice		
Molinate	Rice		Maize	
Pebulate			Maize	
Perfluidone	Maize	Rice	Maize	
Sulfallate			Maize	
Tri-allate		Wheat Oat		Wheat
Vernolate		Maize	Maize	Maize

In our own research (Table 2), R-25788 was found to be
completely inactive as an antidote for EPTC in five different
broadleaved crop and weed species. In all of the research
reviewed by Blair *et al.*, only moderate antidote activity for
NA in cotton and for either NA or R-25788 in field bean was
reported. This would seem to indicate that the antidote action of these two compounds is primarily restricted to grasses and that it may relate to some morphological or physiological characteristic common to many grasses but present in
few broadleaved plants. One might also conclude that NA is
less specific than R-25788 with respect to the herbicides it
can counteract in plants (Table 6). These generalizations
would be more valid if it was possible to know more about
experiments where R-25788 or NA have been found to be inactive
as antidotes for various different plants or various different
herbicides (Tables 2 and 3). Unfortunately, negative results
are seldom reported except as parts of other studies reporting
some type of positive antidote interaction.

V. EFFECT OF R-25788 ON EPTC UPTAKE AND METABOLISM IN CORN

When 10^{-4} M EPTC was applied to corn in quartz sand nutrient culture, severe reduction in corn shoot growth resulted
from only two days exposure (Fig. 5). R-25788 at a concentration of 10^{-5} M was most effective for preventing EPTC

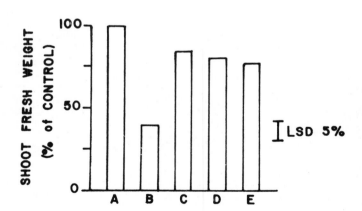

*Fig. 5. Phytotoxicity of 10^{-4} M EPTC to corn under the
influence of 10^{-5} M R-25788 added to the nutrient solution at
different times for a period of 2 days: (A) control; (B)
EPTC alone; (C) R-25788 and EPTC applied at the same time;
(D) R-25788 applied 2 days prior to EPTC; (E) R-25788 applied
2 days after EPTC treatment.*

injury if it was supplied simultaneously with EPTC in the
nutrient solution. However, even if the corn seedlings were
removed after two days exposure to EPTC alone and then treated with R-25788 in the nutrient solution, the EPTC injury
that would have occurred was still significantly reduced. It
can be concluded from this study that R-25788 does not prevent
EPTC injury to corn solely by preventing EPTC uptake.

In other studies (15), the influence of R-25788 on the
uptake and fate of ^{14}C-EPTC in corn was examined. Corn seedlings in the two leaf stage were placed in 50 ml Erlenmeyer
flasks containing 5 x 10^{-5} M ^{14}C-EPTC with or without 10^{-5} M
R-25788 in the nutrient solution. The flasks were sealed
with a plug of parafilm and plasticene around the stem of the
corn seedlings. The flasks and treated corn seedlings were
then placed in closed Mason jars fitted with glass inlet and
outlet tubing. In an adjacent ice bath, two gas washing bot-
tles were connected in series to the "closed system" for each
Mason jar; one containing 0.5 N potassium hydroxide to trap
$^{14}CO_2$ and one containing n-hexane to trap ^{14}C-EPTC or other
metabolites. Filtered, dried air was passed through the system at a rate of 30 ml/min. At the end of the 60 hr treatment period, the nutrient solution, roots, shoots, and trap-
ping solutions were assayed for radioactivity (Table 7).

It is clearly established by these studies that R-25788
does not prevent EPTC injury to corn by preventing EPTC uptake. In fact, in the R-25788 treated plants, greater rates

TABLE 7

*Influence of R-25788 on ^{14}C-EPTC Uptake, Accumulation,
Metabolism to CO_2, or Emanation as a Vapor by Corn
Seedlings*

Radioactivity Recovered as	Total Dpm Recovered After 60 Hr*	
	R-25788 Concn	
	0	10^{-5} M
Treatment solution	724,470	657,500
[^{14}C] EPTC vapor	21,445	32,522
$^{14}CO_2$	10,200	16,097
Plant tissue	151,950	177,061
Unrecoverable	77,935	102,820

* Data represent the average from three experiments. Average
fresh weights of corn plants at harvest time: 2.253 g in
R-25788 treatment and 1.995 g in treatment without R-25788.

of ^{14}C-EPTC disappearance from the solution, accumulation within the plant, metabolism to $^{14}CO_2$, and emanation as unaltered EPTC vapours were observed (Table 7). While the rate of EPTC metabolism was greater in R-25788 treated corn seedlings, the metabolites were not identified and the differences did not seem great enough to compensate for the greater rates of root uptake.

VI. STRUCTURE/ACTIVITY RELATIONSHIPS AMONG ANTIDOTES FOR THIOCARBAMATE HERBICIDES IN CORN

R-25788 and EPTC are very similar in structure (Fig. 6). Both contain an amide group with two N-substituted three-carbon chains. EPTC possesses a sulfur in the acid chain and it is the sulfur which is assumed (16, 18) to be essential for herbicidal activity. In place of the electronegative sulfur in EPTC, R-25788 possesses two electronegative chlorines at nearly the same position in the molecule.

$$CHCl_2\underset{\underset{O}{\|}}{C}-N\begin{smallmatrix} CH_2CH=CH_2 \\ \\ CH_2CH=CH_2 \end{smallmatrix} \qquad CH_3CH_2-S-\underset{\underset{O}{\|}}{C}-N\begin{smallmatrix} CH_2CH_2CH_3 \\ \\ CH_2CH_2CH_3 \end{smallmatrix}$$

R-25788   EPTC

Fig. 6. A comparison of molecular structures for R-25788 (antidote) and EPTC (herbicide).

To examine the significance of the similarities between these molecules, over 25 analogues of R-25788 were synthesized and compared for activity as antidotes for EPTC (19). Related series of antidotes were synthesized to determine the most active chain length and degrees of unsaturation in the N-substituted alkyl groups. Various lengths of acid chains and the importance of chlorines in the acid chain were also examined.

In the original patent for R-25788 (7) and in other investigations (17, 18), many analogues of R-25788 have been examined as antidotes for EPTC in corn. However, in these situations, antidote activity was compared in soil bioassay systems and the rates of application often resulted in complete prevention of EPTC injury in corn for most compounds. Thus few structure/activity differences were evident and even

those that were apparent could have related to differential
interactions in the soil rather than to differences in activity within the plant. To compare differences in activity
that were strictly "plant related", a soil-free bioassay system in quartz sand nutrient culture was employed. Corn seeds
(United Hybrid 106, Stewart 2501, Golden Beaver) were germinated on filter papers in petri dishes for two days at 24°C.
When the radicles were 1.5-2 cm in length, the seedlings were
transplanted at a depth of 2 cm in 9 x 7 cm styrofoam cups
containing quartz sand. The bottom of the cup containing the
sand was perforated with 16-20 small holes and then placed in
a lower cup of the same size to which was added 40 ml of the
half strength Hoagland's nutrient solution. The plants were
maintained in growth rooms with a 16 hr light period (4600
lux light intensity) at 24°C and 20°C during the 8 hr dark
period.

Two days after transplanting when the corn seedlings
were in the two-leaf stage and approximately 6 cm in height,
the initial nutrient solution was drained out and the treatments applied. EPTC and antidote solutions were added simultaneously in 20 ml volumes to provide for the following treatments in 40 ml of nutrient solution: untreated control, EPTC
10^{-4} M, antidote 10^{-4} M, and EPTC 10^{-4} M plus antidote 10^{-4} M.
At 4 days after transplanting, freshly prepared treatment
solutions were reapplied. These concentrations were chosen
on the basis of preliminary studies in which two applications
of 10^{-4} M EPTC caused a severe reduction in corn shoot growth
that was seldom completely overcome by 10^{-4} M concentrations
of the most active antidote analogues. Preliminary bioassays
of all the synthesized analogues established that N-monosub-
stituted dichloroacetamides (N-allyl-2,2-dichloroacetamide)
were much lower in activity as EPTC antidotes than equivalent
N,N-disubstituted dichloroacetamides (R-25788). Branched acid
chains [$(CH_3)_2CHCON(CH_2CH=CH_2)_2$] also had little or no activity as antidotes. Pallos *et al.* had observed similar com-
parisons between N-mono and N,N-disubstituted acetamides with
bioassays in soil (17).

Comparative studies with dichloroacetamides which varied
in length of N,N-substituents established that N,N-di-n-
propyl-2,2-dichloroacetamide was the most active as an antidote for EPTC (Table 8). Similar analogues with N,N-substi-
tuted three carbon chains containing double or triple bonds
were also less active than N,N-di-n-propyl 2,2-dichloroace-
tamide (Table 8).

With N,N-diallyl acetamides (Table 9), either one or two
chlorines were required in the acid chain for optimum activity

in preventing EPTC injury in corn. In contrast to this, in the $\underline{N},\underline{N}$-dipropyl series of analogues, two chlorines were clearly essential in the acid chain for greatest antidote activity (Table 9).

TABLE 8

Comparative Activity of Dichloroacetamides Varying in Length and Saturation of Amide Chains as Antidotes to EPTC in Corn Grown in Quartz Sand Nutrient Culture

Antidotes Added to EPTC*

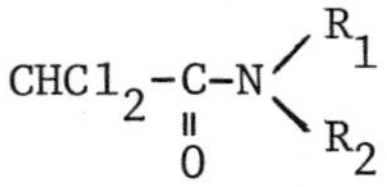

$$CHCl_2-\overset{\parallel}{\underset{O}{C}}-N\overset{R_1}{\underset{R_2}{<}}$$

Amide Chains		Corn Injury (% Reduction in Shoot Dry Weight)
R_1	R_2	
Saturation		
$CH_3CH_2CH_2-$	$CH_3CH_2CH_2-$	5 d
$CH_2=CHCH_2-$	$CH_2=CHCH_2-$	14 c
$CH\equiv CH-CH_2$	$CH\equiv CHCH_2-$	26 b
Chain length		
CH_3CH_2-	CH_3CH_2-	46 a
$CH_3CH_2CH_2-$	$CH_3CH_2CH_2-$	5 d
$CH_3CH_2CH_2CH_2-$	$CH_3CH_2CH_2CH_2-$	31 b
EPTC alone (control)		45 a

* EPTC and antidotes were applied simultaneously at equimolar concentrations of 10^{-4} M in quartz sand nutrient culture to 4 day germinated corn seedlings. Plants were harvested 11 days after treatment. Means followed by unlike letters significantly differ (.05 level, ANOV and Duncan's).

TABLE 9

Influence of Degree of Acid Chain Chlorination on the Activity of N,N-di-n-propyl and N,N-diallyl acetamides as Antidotes for EPTC in Corn Grown in Quartz Sand Nutrient Culture

---

ANTIDOTES ADDED WITH EPTC
N,N-Diallyl Acetamides

$$R-N\begin{matrix} CH_2CH{=}CH_2 \\ CH_2CH{=}CH_2 \end{matrix}$$

	Corn Injury (% Reduction in Shoot Dry Weight)
R	
CH_3CO-	51 b
CH_2ClCO- (CDAA)	27 c
$CHCl_2CO-$ (R-25788)	27 c
CCl_3CO-	52 b
EPTC alone	64 a

N,N-Dipropyl Acetamides

$$R-N\begin{matrix} CH_2CH_2CH_3 \\ CH_2CH_2CH_3 \end{matrix}$$

R	
CH_3CO-	51 ab
CH_2ClCO-	43 b
$CHCl_2CO-$	26 c
CCl_3CO-	49 b
EPTC alone	59 a

* EPTC and antidotes were applied at equimolar (10^{-4} M) concentrations to 5 day germinated corn seedlings which were harvested 11 days after treatment. Means (4 replicates) followed by unlike letters significantly differ (.05 level).

N,N-Diallyl amides which varied in numbers of carbons in the acid chain (CH_3CO-, CH_3CH_2CO-, or $CH_3CH_2CH_2CO-$) were only slightly active as antidotes for EPTC (Table 10).

TABLE 10

Comparative Activity of Various N,N-Diallyl Amides with Different Acid Chains as Antidotes for EPTC in Corn Grown in Quartz Sand Nutrient Culture

Antidotes Added to EPTC* $R-N(CH_2CH=CH_2)_2$ R	Corn Injury (% Reduction in Shoot Dry Weight)
CH_3CO-	46 b
CH_3CH_2CO-	43 b
$CH_3CH_2CH_2CO-$	49 ab
$CH_3CH_2-O-CO-$	26 d
EPTC alone (control)	56 a

* Antidotes and EPTC were applied at equimolar (10^{-4} M) concentrations in nutrient solution to 4 day germinated corn seedlings in quartz sand. Plants were harvested 11 days after treatment. Means (4 replicates) followed by unlike letters significantly differ (.05 level).

In one of the analogues in this study (Table 10), an oxygen was included in the acid chain and this resulted in a very significant increase in antidote activity. It is remarkable how much this latter compound resembles EPTC in structure (Fig. 7).

It can be concluded from this series of experiments that molecules very similar to EPTC are very active as antidotes for this herbicide in corn. For good antidote activity there appears to be a requirement for electron withdrawing groups in either or both amide or acid chains. With two chlorines in the acid chain of R-25788 [$CHCl_2CON(CH_2CH=CH_2)_2$], it

N,N-diallyl ethyl
    carbamate

EPTC

Fig. 7. Comparison of molecular structures for N,N-diallyl ethyl carbamate (an active antidote) and EPTC (herbicide).

appears that the electronegative N,N-diallyl groups
(Fig. 6) are not essential for high antidote activity
at least in a soil-free system. With N,N-di-n-propyl acetamide, two chlorines are required in the acid chain for the
molecule to be sufficiently electronegative. The high activity of diallyl ethyl carbamate as an antidote establishes that
chlorine(s) or an oxygen can be employed in the acid chain.
It seems possible that the chlorines or the oxygen result in
acid chains that are similarly electronegative but less toxic
than the *S*-ethyl group in the thiocarbamate linkage of EPTC.

To further examine the idea that molecules very similar
to thiocarbamates can be active antidotes for these herbicides in corn, dichloroacetamides were synthesized with N,N-disubstituted alkyl chains identical to each of four other
thiocarbamate herbicides; EPTC, vernolate, pebulate, molinate,
and butylate (Table 11). These five synthesized dichloroacetamides were then compared for activity as antidotes for each
of the five thiocarbamate herbicides. While the concentration
of the herbicides was varied so that corn shoot growth would
be reduced to approximately 50% of the control or less, all
of the antidote analogues were evaluated at 10^{-4} M in the
nutrient solution.

R-25788 was moderately active as an antidote for all of
these thiocarbamate herbicides and highly active for reducing
corn injury from EPTC or vernolate (Table 12). N,N-Di-n-propyl-2,2-dichloroacetamide was at least as active as
R-25788 as an antidote for all five of these herbicides.
Most significant was the fact that for each of the thiocarbamate herbicides, the most active antidote was the compound
most similar to it with respect to the N,N-disubstituted
alkyl groups. N,N-Di-n-propyl-2,2-dichloroacetamide was significantly the most active for preventing EPTC or vernolate

TABLE 11

Structures of Five Dichloroacetamides Synthesized for Examination as Antidotes for Five Thiocarbamate Herbicides with Similar Amide Chains

Herbicides	Structure	Corresponding Dichloroacetamides
EPTC	$CH_3CH_2\text{-}S\text{-}\overset{\displaystyle O}{\overset{\|}{C}}\text{-}N(CH_2CH_2CH_3)_2$	$CHCl_2\text{-}\overset{\displaystyle O}{\overset{\|}{C}}\text{-}N(CH_2CH=CH_2)_2$
vernolate	$CH_3CH_2CH_2\text{-}S\text{-}\overset{\displaystyle O}{\overset{\|}{C}}\text{-}N(CH_2CH_2CH_3)_2$	$CHCl_2\overset{\displaystyle O}{\overset{\|}{C}}\text{-}N(CH_2CH_2CH_3)_2$
pebulate	$CH_3CH_2CH_2\text{-}S\text{-}\overset{\displaystyle O}{\overset{\|}{C}}\text{-}N\begin{smallmatrix} CH_2CH_3 \\ \diagdown \\ CH_2CH_2CH_2CH_3 \end{smallmatrix}$	$CHCl_2\overset{\displaystyle O}{\overset{\|}{C}}\text{-}N\begin{smallmatrix} CH_2CH_3 \\ \diagdown \\ CH_2CH_2CH_2CH_3 \end{smallmatrix}$
molinate	$CH_3\text{-}S\text{-}\overset{\displaystyle O}{\overset{\|}{C}}\text{-}N\begin{smallmatrix} CH_2CH_2CH_2 \\ \diagdown \quad\| \\ CH_2CH_2CH_2 \end{smallmatrix}$	$CHCl_2\overset{\displaystyle O}{\overset{\|}{C}}\text{-}N\begin{smallmatrix} CH_2CH_2CH_2 \\ \diagdown \quad\| \\ CH_2CH_2CH_2 \end{smallmatrix}$
butylate	$CH_3CH_2\text{-}S\text{-}\overset{\displaystyle O}{\overset{\|}{C}}\text{-}N[CH_2CH(CH_3)_2]_2$	$CHCl_2\overset{\displaystyle O}{\overset{\|}{C}}\text{-}N[CH_2CH(CH_3)_2]_2$

54

TABLE 12

Comparative Activity of 5 $\underline{N},\underline{N}$-Disubstituted 2,2-Dichloroacetamides as Antidotes for 5 Different Thiocarbamate Herbicides in Corn Grown in Quartz Sand Nutrient Culture*

Antidotes Added With Thiocarbamate Herbicides (10^{-4} M in Nutrient Culture) $$\underset{\mid}{\overset{O}{\underset{\mid}{\overset{\parallel}{CHCl_2-C-N}}}}\overset{R_1}{\underset{R_2}{\Big\langle}}$$		Corn Injury (% Reduction in Shoot Dry Weight) Conc. (x 10^{-4} M) in Nutrient Culture				
R_1	R_2	EPTC 1.0	Vernolate 1.3	Pebulate 1.3	Molinate 2.2	Butylate 10.3
$CH_3CH_2CH_2-$	$CH_3CH_2CH_2-$	4 d	20 d	42 c	53 c	65 ab
$CH_2=CHCH_2-$	$CH_2=CHCH_2-$	31 c	26 cd	44 c	54 c	61 b
CH_3CH_2-	$CH_3CH_2CH_2CH_2-$	55 a	28 c	37 c	47 d	47 c
$CH_3\overset{CH_3}{\underset{\mid}{CH}}CH_2-$	$CH_3\overset{CH_3}{\underset{\mid}{CH}}CH_2-$	55 a	46 b	71 a	60 b	40 c
$\overset{CH_2CH_2CH_2}{\underset{CH_2CH_2CH_2}{\Big\rangle}}$		47 b	60 a	52 b	46 d	72 a
Control (respective herbicides alone)		45 b	56 a	77 a	74 a	69 a

* Within columns for each herbicide, means followed by unlike letters significantly differ (.05 level).

55

injury while R-25788 with N-substituted allyl groups was the
next most active. For molinate, butylate, or pebulate, the
corresponding N,N-disubstituted acetamide was mathematically
the most effective antidote and was statistically in the group
of the two or three most active analogues. The cyclic antidote was most active for molinate and least active for butylate. Conversely the N,N-diisobutyl antidote was most active
for butylate and least active for molinate.

It would be incorrect to assume that identical structure/
activity relationships to those obtained here could also be
observed in bioassays in soil or under actual field use situations. Differences in soil-sorption interactions or in
antidote degradation rates in soil could certainly be expected to alter comparative activity relationships.

It would also be incorrect to conclude that close structural similarities to thiocarbamates are requirements for
molecules to have significant activity as antidotes for these
herbicides in corn. Certainly there is ample evidence in the
patent for R-25788, where over 500 different amides were examined, that quite different molecules can have activity as
antidotes (7). It is possible that all of these antidotes
may have a similar mode of action and that, in a field use
situation in soil, availability for uptake by the plant is
the most important criterion. It is also possible that antidotes markedly different in structure could have different
modes of action from antidotes that closely resemble the
herbicide in structure. However, it is clearly established
in this study that highly similar antidote molecules do have
high activity as antidotes for thiocarbamate injury in corn.
Since these correlations were observed with a soil free bioassay system, it is apparent that they relate to differences
in activity within the plant and should therefore be highly
relevant to the mechanism involved. With the bioassays in
soil employed by other investigators, this aspect has most
likely not been evident and has not been previously empha-
sized.

VII. DISCUSSION

The development of antidotes to selectively reduce herbicide injury in crops is a new and exciting area and the
growing interest in this field is not surprising. The antidote approach may well offer the best opportunity to develop
control methods for serious weeds in major crops with our
current spectrum of available herbicides. In their efforts
to capitalize on herbicide-antidote technology, some investi-

gators will undoubtedly embark on screening programs involving most combinations of important herbicides, possible antidotes, and major crops. This empirical approach will certainly result in some rewarding successes. Nevertheless, a more systematic approach using current knowledge of antidote action could be more efficient.

On the basis of results reported earlier in this paper, it seems logical to conclude that NA and R-25788 act by different mechanisms as antidotes for EPTC since R-25788 is highly specific for protecting only corn while NA is not. On the basis of present evidence, it can also be concluded that R-25788 (Fig. 5, Table 7) and NA (20, 21) do not prevent herbicide injury to plants by preventing herbicide uptake. It seems more probable that the antidotes either prevent activation or enhance deactivation of the herbicides within the plant. Independent effects of the antidotes and the herbicides that counteract each other within the plants are also possible.

Holm and Szabo (21) and Murphy (20) reported that NA seed treatment increased the rates of herbicide metabolism in protected plants. Lay and Casida (16, 18) reported that EPTC in corn was converted to a more toxic EPTC-sulfoxide metabolite. They showed that R-25788 induced higher levels of glutathione and glutathione-$\underline{S}$-transferase sufficient to prevent toxic accumulation of the EPTC-sulfoxide by carbamoylation. This mechanism might be the correct explanation for the EPTC-chloroacetamide interactions in corn. It is important, however, to determine how widely this explanation is correct. Are all of the herbicides (Table 6) for which R-25788 is effective as an antidote detoxified by conjugation with glutathione? According to Lay and Casida (16), corn normally has high levels of glutathione and further increases by R-25788 result in sufficient amounts to detoxify the EPTC-sulfoxide, if present. In other plants, like oats, where glutathione levels are normally low, slight increases with R-25788 are not sufficient to prevent EPTC-sulfoxide toxicity. It would follow then that if there are plants other than corn that are also normally high in glutathione content, these plants should also be protected from EPTC by R-25788. Also, R-25788 should be effective as an antidote for other herbicides such as various s-triazines (22) that are also known to be detoxified in plants by glutathione conjugation.

Wilkinson and coworkers (23, 24, 25) have reported on several effects of thiocarbamate herbicides in broadleaved plants, some of which can be influenced by antidotes. EPTC can reduce the deposition of cuticular waxes on plant leaf

surfaces (23) and inhibit fatty acid synthesis in spinach
(*Spinacea oleracea* L.) chloroplasts (24) and in red beet
(*Beta vulgaris* L.) root sections (25).   EPTC also increased
the efflux of betacyanin from red beet roots (25).   Even
though NA and R-25788 can counteract the effects of EPTC on
fatty acid synthesis, it is difficult to interpret these
effects in relation to the possible primary mechanisms for
herbicidal action.   Can these same effects of EPTC be observed in grasses?   R-25788 does not seem to be an effective
antidote for EPTC in any plant except corn.   How, then, can we
interpret the significance of EPTC effects in broadleaved
plants that can be clearly antidoted by R-25788?   It is possible that EPTC has many effects in plants and that its primary toxic effect is normally not operative in broadleaved
plants making the less toxic secondary effects on cuticle
deposition and fatty acid synthesis more readily observed.

In our own research, we have shown that compounds structurally very similar to the thiocarbamates can be highly
active as antidotes for these herbicides in corn.   With our
present knowledge, the significance of these findings is also
difficult to assess.   Some of our data would certainly support the theory that the antidote acts as a competitive inhibitor for EPTC within the plant.   A chemically similar antidote could possibly compete with EPTC for the site on the
enzyme which produces the toxic EPTC-sulfoxide.   It could
also compete with the sulfoxide for some toxic site of action.
If this competitive inhibition theory is correct we would
expect a single antidote compound with the highest affi-
nity for the site to be the most active antidote for all thiocarbamate herbicides, however this was not found to be true
(Table 12).   These results could indicate that the various
thiocarbamate herbicides have slightly different mechanisms
and sites of action that can be antidoted by different competitive inhibitors for each.   While competitive inhibition
might explain the activity of chemically similar compounds
as antidotes, it clearly does not explain how structurally
unrelated compounds can be active as antidote treatments.
With our present knowledge, we can only suggest that the high
activity of chemically similar molecules is more than a coin-
cidence and that in the search for antidotes to other herbicides, examining structurally similar compounds for antidote
activity could be a rewarding first approach.

It may be unrealistic to expect a single well
defined mechanism for compounds that act as herbicide antidotes.   This is particularly true for the examples currently
under study since the herbicides involved seem to have multiple and varied effects on plants.   It may be helpful to

indicate the properties which many of these herbicides (Table 6) have in common. 1) They are usually more toxic to grasses than to broadleaved plants, 2) uptake through shoots of grass seedlings, either from soil or foliar applications, is more effective than root uptake, 3) the site of action is most likely at or near the coleoptillar node, 4) they are not acutely toxic to plants, not even to grasses, and 5) they seem to slow down and eventually stop grass shoot growth, producing plants with some of the following symptoms: twisted, bent, tightly rolled, leathery, brittle, or dark green leaves.

Some investigators (26) have found EPTC injured corn seedlings to have symptoms very similar to those observed in corn treated with classical growth retardants or to those observed in dwarf corn mutants. It is possible that many of these herbicides have effects at the hormonal level within the meristematic regions of grass shoots. For example, Chen *et al.* (27) reported that molinate inhibition of s-RNA synthesis could be reversed with gibberellic acid (GA). Best and Schrieber (28) reported that three out of four types of RNA syntheses inhibited by EPTC in corn could be reversed by treatment with 2,4-D. Harvey *et al.* (26) reported that the stunting effect of EPTC on corn shoots could be prevented by exogenous applications of GA but that the leaf twisting and rolling effects of EPTC were not affected. Several investigators have reported that NA plus GA is more effective for preventing herbicide injury to grasses than the NA seed treatment alone (29, 30, 31). As with other areas of antidote research, the significance of these hormonal interactions is not fully understood.

Our present state of knowledge of herbicide antidotes indicates that there are many uncertainties. We seem to be only approaching the threshold of understanding. Further research could well yield proof for some of the theories of antidote action outlined above. Certainly it is important to establish which effects of these herbicides and antidotes are of primary or secondary importance. Hopefully those researchers who attempt to clarify some of the uncertainties in this area will be rewarded with yet further discoveries of how this approach to selective weed control may be applied.

VIII.   ACKNOWLEDGEMENTS

These investigations have been funded in part by grants to the senior author from the National Research Council of Canada. The authors wish to thank J.D. Bandeen, G.W. Anderson, and N.J. Bunce for consultation and collaboration in these investigations and J. McLeod, R. Makowski, J. Curry, E. Pelly, L. Barron, and S. Ball for their technical assistance.

IX. REFERENCES

1. Truffaut, C., and Pastac, I., British Patent 425,295 (1935).
2. Slade, R.E., Templeman, W.G., and Sexton, W.A., Nature 155, 497 (1971).
3. Marth, P.C., and Mitchell, J.W., Bot. Gaz. 106, 224 (1944).
4. Ashton, F.M., and Crafts, A.S., "Mode of Action of Herbicides", p. 20. Wiley-Interscience, New York, 1973.
5. Hoffmann, O.L., Weeds 10, 322 (1962).
6. Hoffmann, O.L., Weed Sci. Soc. Amer. Abstracts 12 (1969).
7. Pallos, F.M., Brokke, M.E., and Arneklev, D.R., Stauffer Chemical Co., Belgian Patent 782,120   (1972).
8. Bandeen, J.D., Can. Weed Comm., Res. Rep. East. Sec. 20 (1971).
9. Brown, R.H., and Shaw, J.E., Can. Weed Comm., Res. Rep. East. Sec. 27 (1971).
10. Chang, F.Y., Bandeen, J.D., and Stephenson, G.R., Can. J. Plant Sci. 52, 707 (1972).
11. Chang, F.Y., Stephenson, G.R., and Bandeen, J.D., Weed Sci. 21, 292 (1973).
12. Chang, F.Y., Bandeen, J.D., and Stephenson, G.R., Weed Res. 13, 399 (1973).
13. Chang, F.Y., Stephenson, G.R., Anderson, G.W., and Bandeen, J.D., Weed Sci. 22, 546 (1974).
14. Blair, A.M., Parker, C., and Kasasian, L., PANS 22, 65 (1976).
15. Chang, F.Y., Stephenson, G.R., and Bandeen, J.D., J. Agr. Food Chem. 22, 245 (1974).
16. Lay, M.M., and Casida, J.E., Science 189, 287 (1975).
17. Pallos, F.M., Gray, R.A., Arneklev, D.R., and Brokke, M.E., J. Agr. Food Chem. 23, 821 (1975).
18. Lay, M.M., and Casida, J.E., Pest. Biochem. Physiol. 6, 442 (1976).
19. Stephenson, G.R., Bunce, N.J., Makowski, R.I., and Curry, J.C., J. Agr. Food Chem. 24, in press (1977).
20. Murphy, J.J., Chem. Biol. Int. 5, 284 (1972).
21. Holm, R.E., and Szabo, S.S., Weed Res. 14, 119 (1974).
22. Lamoureux, G.L., Shimabukuro, R.H., Swanson, H.R., and Frear, D.S., J. Agr. Food Chem. 18, 81 (1970).
23. Wilkinson, R.E., Plant Physiol. 53, 269 (1974).
24. Wilkinson, R.E., and Smith, A.E., Weed Sci. 23, 90 (1975).
25. Wilkinson, R.E., and Smith, A.E., Weed Sci. 24, 235 (1976).
26. Harvey, B.M.R., Chang, F.Y., and Fletcher, R.A., Can. J.

Bot. 53, 225 (1975).

27. Chen, T.M., Seaman, D.E., and Ashton, F.M., Weed Sci. 16, 28 (1968).
28. Best, C.E., and Schrieber, M.M., Weed Sci. 20, 4 (1972).
29. Guneyli, E., Diss. Abst. Int. B 32(4), 1957 (1971).
30. Hahn, R.R., Diss. Abst. Int. B 35(4), 1483 (1974).
31. Parker, C., and Dean, M.C., Pestic. Sci. 7, 403 (1976).

FIELD APPLICATIONS OF THIOCARBAMATE ANTIDOTES

F. W. Slife
University of Illinois

The incidence of corn injury from thiocarbamate herbicides is effectively reduced by seed treatment with 1,8-naphthalic anhydride and seed or soil treatment with the dichloroacetamide R-25788. Additional selectivity provided by this dichloroacetamide expands the use areas for EPTC and butylate in weed control.

I. INTRODUCTION

From the standpoint of a producer there are two major requirements for a herbicide to be used widely in crop production.

1. It must kill weeds consistently.

2. It must have crop tolerance.

Crop tolerance to most selective herbicides has been and continues to be a problem. Even though field rates are adjusted to all of the parameters that are known to effect crop tolerance, crop injury may result. Atrazine is the exception since corn possesses so much tolerance that injury rarely occurs.

Thiocarbamates as a general class of herbicides are considered to possess only fair crop tolerance although there are some exceptions. Apparently many potential thiocarbamate herbicides have been discarded because of inadequate crop tolerance. This has been disappointing because the spectrum of weed control has been good and in general they have relatively short soil life thus avoiding residues that may interfere with succeeding crops.

II.   THIOCARBAMATE ANTIDOTES

Herbicide antidotes are a major contribution to increasing the selectivity of thiocarbamate herbicides and thus reducing the incidence of crop injury.

Activated charcoal was perhaps the first material used as a herbicide antidote. It apparently was first employed to inactivate excess quantities of herbicides in soils, but later it was successfully used as a seed treatment and as a root drench to reduce herbicide injury. Hoffman (1) in 1962 introduced the antidote concept and reported on a number of compounds that were possibilities for herbicide antidotes. In the late 1960's, as a result of Hoffman's work, Gulf Chemical introduced 1,8-naphthalic anhydride as a seed protectant. Using this material as a corn seed treatment, Burnside et al. (2) at Nebraska reported a marked reduction in corn injury where the thiocarbamate herbicides butylate and EPTC were used. Field tests at 3 locations for 2 years indicated that 10 kg/ha of EPTC reduced yields 50% without anhydride, but where the corn seed was treated with anhydride yields were comparable to cultivated check plots.

In 1970 Stauffer Chemical introduced N,N-diallyl-2,2-dichloroacetamide (R-25788) as an antidote to thiocarbamate injury. This material gave results similar to anhydride when used as a seed treatment, but it possessed the added flexibility that it could be included in the herbicide formulation and thus applied with the herbicide treatment (3). It has been tested with a variety of thiocarbamate herbicides with positive results.

Laboratory and field work in 1970 and 1971 (4, 5) clearly demonstrated that even though corn tolerance was improved greatly with R-25788, it did not change the toxicity of the thiocarbamate herbicide to sensitive annual weed species.

The success of adding an antidote to butylate and EPTC is evident from the corn acreage treated with these compounds. Treated acreage is now in the neighborhood of 20% of the total U. S. corn acreage. These two thiocarbamates are excellent annual grass control materials for corn and in addition their use has been expanded to control other serious weed problems. Johnsongrass (Sorghum halepense) and shatter cane (Sorghum bicolor) are weedy sorghum species that were not well controlled in land where corn was grown. Both butylate and EPTC have given substantial relief to these problems. Yellow nutsedge (Cyprus esculentus) is spreading rapidly in the corn belt area and again EPTC with the antidote has proven to be a very acceptable control material in corn.

Corn breeders and geneticists have been successful in eliminating some of the germ plasm that is extremely sensitive to herbicide injury (6). Variation still exists but much improvement has been made. With introduction and use of antidotes, variation in reaction to thiocarbamate herbicides is much reduced and it is now possible to utilize some high yielding inbreds that were extremely sensitive to thiocarbamate prior to the introduction of the antidote.

The present diallyl dichloroacetamide antidote is not perfect in that under some soil conditions it appears to degrade at a more rapid rate than the herbicide. In these cases herbicide injury can be expressed at a later stage of corn growth. Fortunately, these cases are relatively rare.

The commercial use of an antidote with the thiocarbamate herbicides butylate and EPTC has made a major contribution to weed control. Because of increased selectivity these herbicides are now major use compounds. The additional selectivity has allowed them to be used at higher rates for the control of certain perennial weeds.

III.   REFERENCES

1.   Hoffman, O. T., Weeds 10, 322-323 (1962).
2.   Burnside, O. C., G. A. Wicks , and C. R. Fenster,
         Weed Sci. 19, 565-568 (1971).
3.   Chang, F. Y., G. R. Stephenson, and J. D. Bandeen,
         Weed Sci. 21, 292-295 (1973).
4.   Burt, G. W., Weed Sci. 24, 319-321 (1976).
5.   Burt, G. W., Weed Sci. 24, 327-330 (1976).
6.   Wright, T. H., and C. E. Rieck, Weed Sci. 22, 83-85
         (1974).

Part IV

Physiological Actions of Thiocarbamate Herbicides and Their Antidotes

SITE OF UPTAKE AND ACTION OF THIOCARBAMATE HERBICIDES AND HERBICIDE ANTIDOTES IN CORN SEEDLINGS

Reed A. Gray and Grant K. Joo
Stauffer Chemical Company

Exposing different parts of corn seedlings to vapors of the herbicide, S-ethyl dipropylthiocarbamate (EPTC), and the antidote, N,N-diallyl-2,2-dichloroacetamide (R-25788), showed that the aboveground parts of the shoots were ineffective sites of uptake for inducing injury from the herbicide or protection with the antidote. However, the underground parts of the shoots and the roots were effective sites for uptake of the herbicide and antidote when applied as vapors for several days and when applied to the soil using charcoal barriers to protect the untreated parts. When EPTC was applied to the soil in the root zone, injury to the corn resulted and the best protection was obtained when the antidote was also applied to the root zone. More severe injury occurred when the herbicide was applied to the soil in the shoot zone, and R-25788 offered better protection when applied to the soil in the same shoot zone than when applied to the roots. Application of EPTC and R-25788 in lanolin paste to different parts of corn seedlings each gave maximum effectiveness when applied in a band around the corn shoot at a point 0-5 mm above the coleoptilar node located about 1 cm below the soil surface. This indicated that the meristematic region of the corn shoot is the site of action of both the herbicide and the antidote.

I. INTRODUCTION

As an aid to mode of action studies on herbicide antidotes and in order to obtain a better understanding of how they perform, it was of interest to determine the sites of uptake and sites within the corn plant where the antidotes exert their protective action. In order to facilitate these studies with the antidotes, more information was needed on the sites of uptake and action of the thiocarbamate herbicides in corn seedlings. Therefore, site of uptake and action studies were carried out concurrently with both the herbicide, S-ethyl dipropylthiocarbamate (EPTC), and the antidote, N,N-diallyl-2,2-dichloroacetamide (R-25788), in corn seedlings. Several different methods of applying the herbicide and the antidote to different parts of the corn seedlings were investigated including application by the following methods: as vapors in a closed jar, by the split-root tech-

nique, to the soil in the root or shoot zone, by foliage dips
and sprays, and in lanolin paste. The charcoal barrier technique developed previously (1) was used to prevent movement
of the vapors and solutions in the soil from one zone to
another, so that certain parts of the seedlings could be
treated without exposing the other parts.

A number of reports (1-7) have been concerned with the
site of uptake of EPTC in various plant species. In some
species root exposure to the treated soil gave more injury
than shoot exposure and in other species shoot exposure gave
more injury (1, 5). Corn was injured slightly more by shoot
exposure than root exposure to EPTC using the charcoal barrier
method (1). In corn the shoot area adjacent to the crown node
was extremely sensitive to treatment by EPTC (7).

Very little work has been reported on the site of uptake
of herbicide antidotes, although the first antidotes and
1,8-naphthalic anhydride were most effective when applied to
the seeds (8-10). R-25788 was effective in protecting corn
from injury by EPTC and other herbicides when applied both as
a seed treatment (11) and when mixed into the soil with the
herbicides (10-11). Application of R-25788 in nutrient solution also protected corn from injury by EPTC (12). In these
previous studies where seeds, soil and the nutrient solution
were treated with antidotes, no effort was made to determine
the site of uptake and action by exposing or applying the
antidotes to certain parts of the corn seedlings while protecting the other parts. Therefore such tests were carried
out in the present investigation.

II. APPLICATION AS VAPORS

In a large 28.5-liter glass battery jar were placed four
quart mason jars, each containing four 7-day-old corn (DeKalb
XL-44) seedlings that had been planted 2.5 cm deep in a loamy
sand soil. Just before placing in the vapor chamber, the
upper 2.5 cm of soil was removed from each quart jar, and in
three jars the soil was replaced with a 3 mm layer of a 1:1
mixture of activated charcoal (Darco G-60) and soil applied as
a slurry in one part of water (Figure 1). The top of the
charcoal layer was just below the coleoptilar node. In the
fourth jar, the removed soil was replaced with 3 mm of soil.
A moist filter paper (12.5 cm) was placed on the mouth of an
empty quart jar located in the middle of the four jars containing the corn plants. Then 0.1 ml (96 mg) of technical
EPTC (99.7% purity) was placed in the middle of the filter
paper and the battery jar was sealed with a glass desiccator
lid.

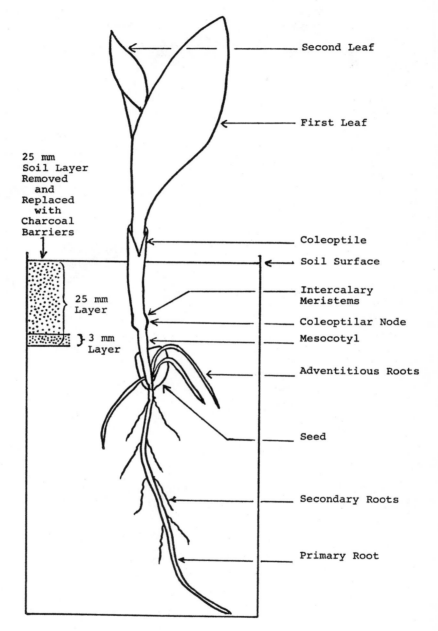

Fig. 1. Seven-day-old corn plant showing the location of the 25 mm layer of soil which was removed and replaced with a 25 mm or 3 mm layer of charcoal in some tests.

One quart jar containing the charcoal barrier and one with no charcoal were removed after 24 hours of vapor treatment and placed in the greenhouse. The large battery jar was flushed with air and the filter paper replaced with another one containing 0.1 ml of fresh EPTC. After 48 hours another jar of plants was removed and the flushing process repeated. After 72 hours a third jar was removed and placed in the greenhouse.

At the same time as the above plants were being treated, an equal number of jars of plants in a second large battery jar were treated in the same way, except that the moist filter paper was treated with 0.1 ml of technical EPTC on the left half and 0.1 ml of technical R-25788 on the right half. If the chemicals vaporized completely this would give concentrations of 3.4 ppm (w/v) of EPTC and 4.2 ppm (w/v) of R-25788 in the air inside the battery jar, so it was assumed that these were the maximum concentrations reached. The battery jars were kept in the greenhouse where the temperature inside the jar ranged from 32 to 37.5°C. The number of plants that developed typical systemic thiocarbamate shoot-twisting or leaf-rolling symptoms was recorded at weekly intervals. The results are shown in Table 1.

TABLE 1

Effect of exposing the shoots of corn seedlings to EPTC and R-25788 vapors at 32-37.5°C using charcoal barriers to shield the roots

Vapor Treatment		Width of Charcoal Barrier Above Seed	Exposure Time	% of Plants* with Leaf-Roll Injury Symptoms	
EPTC	R-25788			8 Days	74 Days
ppm	ppm	mm	Hours	Days	Days
3.4	0	none	24	100	100
3.4	4.2	none	24	0	0
3.4	0	3	24	0	0
3.4	0	3	48	100	0
3.4	0	3	72	100	75
3.4	4.2	3	24	0	0
3.4	4.2	3	48	0	0
3.4	4.2	3	72	66	33

* 4 plants were used for each treatment.

The results in Table 1 show that when the roots were not protected with a charcoal barrier, the EPTC vapor injured all of the plants, but when the R-25788 vapor was also present the plants were protected. When the 3 mm charcoal barrier was present, exposing the corn shoots to EPTC vapor for 24 hours did not produce the typical systemic (leaf-rolling and shoot-twisting) injury symptoms in the corn plants. Some contact injury on the leaves in the form of scalding and yellowing occurred. Exposure for 48 hours and 72 hours produced typical injury symptoms. The plants recovered from the 48-hour treatment after 74 days but not from the 72-hour exposure to EPTC. When the antidote (R-25788) vapor was present with the EPTC vapor for the 48-hour exposure, no injury developed. However, after exposure for 72 hours, the antidote vapor protected only one of the three plants present at 8 days and two of the plants at 74 days.

Another experiment was carried out in the same way as the previous one, except that the vapor chambers were placed in the laboratory at a cooler temperature of 20-25°C. This was done to avoid some of the local leaf-scalding contact injury which occurred in the greenhouse where the sunlight caused heating inside the chambers. In this test, one additional jar of corn seedlings was included in each large battery jar. In this jar the upper 25 mm of soil that was removed was replaced with a 25 mm layer of the charcoal-soil mixture so that none of the underground parts of the corn shoots would be exposed to the chemicals (Figure 1). This jar was removed after exposure to the vapors for 48 hours. The results are shown in Table 2.

The results in Table 2 show again that when no charcoal was present, exposing the corn seedlings to EPTC vapor injured all the plants, and the presence of the R-25788 vapor protected all of the plants. The injury by EPTC vapor alone was attributed mainly to penetration of the vapor into the soil and absorption by the roots, since no injury occurred with the same treatment when the roots and underground parts of the shoots were shielded to a depth of 25 mm with a layer of charcoal. When only the roots were shielded with a 3 mm layer of charcoal, 25% of the plants developed injury after exposure to EPTC vapor for 48 hours, and none were injured after the 24-hour exposure, while 100% were injured after exposure for 72 hours. The presence of the antidote vapor along with the EPTC vapor protected all of the plants exposed for 48 hours and 50% of the plants exposed for 72 hours to the EPTC vapor.

TABLE 2

Effect of exposing the shoots of corn seedlings to
EPTC   and R-25788 vapors at 20-25°C using charcoal
barriers to shield the roots and underground parts
of the shoots

Vapor Treatment		Width of Charcoal Barrier Above Seed	Exposure Time	% of Plants* with Leaf-Roll Injury Symptoms
EPTC ppm	R-25788 ppm	mm	Hours	After 14 days
3.4	0	none	48	100
3.4	4.2	none	48	0
3.4	0	25	48	0
3.4	4.2	25	48	0
3.4	0	3	24	0
3.4	0	3	48	25
3.4	0	3	72	100
3.4	4.2	3	24	0
3.4	4.2	3	48	0
3.4	4.2	3	72	50

*   4 plants were used for each treatment.

A third experiment was carried out in which EPTC was
applied to the roots and R-25788 was applied as a vapor to the
shoots using plastic tubs 15 cm in diameter and 17.5 cm deep
instead of the glass jars. The upper 2.5 cm of soil was removed from each of two pots containing thirteen 7-day-old
corn (DeKalb XL-44) seedlings. The soil was replaced with a
25 mm layer of charcoal in one pot and with a 3 mm layer of
charcoal in the other (Figure 1). The 3 mm layer of charcoal
above the seed protected the roots but allowed exposure of the
underground parts of the shoots to antidote vapor. The 25 mm
layer of charcoal allowed only exposure of the aboveground
parts of the corn shoots. Twenty ml of solution containing
24 mg of EPTC were injected into the soil in the root zone
below the charcoal layer in each pot with a pipette. The hole
made by the pipette was filled with charcoal. After diffusion
in the soil, this would give a concentration of 12 ppm EPTC.
Both pots were then placed in the 28.5-liter glass battery jar
and exposed for 24 hours to the vapor from 0.1 ml of R-25788
and then placed in the greenhouse. The temperature inside
the jar ranged from 32° to 37°C.

The results taken 8 days after treatment showed that 6 of
the 13 plants in the pot with the 25 mm layer of charcoal developed typical leaf-rolling injury symptoms, but none of the

13 plants in the pot with a 3 mm layer of charcoal were injured. This indicated that the antidote was ineffective when
the leaves and aboveground parts of the shoot were exposed to
R-25788 vapor, but it was effective when the underground parts
of the shoot, including the coleoptilar node and meristematic
region just above the node, were included in the exposure.
The plants were still protected 24 days after treatment, but
not 72 days. Apparently the plants did not absorb enough antidote vapor to last as long as EPTC in the root zone lasted.

    These vapor tests indicated that when no charcoal was
present, the EPTC vapors penetrated the soil and were taken
up by the roots from the soil solution. This absorption by
the roots apparently took place for a period of several days,
since exposure of the underground parts of the shoot showed
that exposure for 2 or 3 days was needed to induce injury.
Exposing the underground parts of the corn shoots for only 24
hours did not lead to injury. Exposing the leaves and aboveground parts of the corn shoots to EPTC vapor for as long as
48 hours produced local scalding and chlorotic areas on the
exposed leaves but did not induce typical systemic injury
symptoms, indicating very little if any translocation of EPTC
downward to the shoot meristems located just below the soil
level occurred. The antidote applied as a vapor offered good
protection when no charcoal barriers were used, indicating
that good protection was obtained when the roots were exposed
to the antidote. Partial protection was obtained when R-25788
was applied as a vapor with EPTC to the underground parts of
the shoots. No protection to corn resulted when the leaves
and other aboveground parts of the corn shoots were exposed to
R-25788 vapor when EPTC was applied to the roots.

III. APPLICATION TO THE SOIL IN THE SHOOT OR ROOT ZONE

    An experiment was designed to determine if exposure of
the underground part of the corn shoot to the antidote R-25788
would protect the plant from injury by EPTC taken up through
the roots and vice versa. It was also of interest to observe
the protection obtained when the herbicide and antidote were
applied together in the root zone or shoot zone.

    In the first test, a charcoal barrier was used to separate
the treated soil in the shoot zone from the treated soil in
the root zone. Twelve DeKalb XL-44 corn seeds were planted
2.5 cm deep in seven plastic pots (15 cm in diameter and 17.5
cm deep), and 5 cm deep in seven other pots. After growing in
the greenhouse for three days the seeds had germinated and the
coleoptiles were 1-10 mm in length and the primary roots were
50-100 mm in length. At this time the soil above the seeds

was removed and a solution of EPTC 6-E in 150 ml of water was
applied as a drench to the roots. In some cases the antidote
was applied with the herbicide for root treatment with both
chemicals. Then 200 ml of a charcoal slurry (1 part Norite,
1 part soil and 1 part water) were added which gave a charcoal
barrier 5-7 mm in depth just above the seed separating the
roots from the shoots. The upper 2.5 or 5 cm of soil that had
been removed was treated with EPTC or R-25788 by mixing in a
small 5-gallon cement mixer. This treated soil was placed on
top of the charcoal barrier for shoot exposure to the chemicals. The plants were watered lightly by sprinkling and
placed in the greenhouse. The number of plants that developed
leaf-rolling and stem-twisting thiocarbamate injury symptoms
was recorded at weekly intervals and the results taken after
four weeks are shown in Table 3.

TABLE 3

*Effect of root vs. shoot exposure to R-25788-and EPTC-
treated soil on the amount of corn injury obtained
using charcoal barriers*

Root Zone Treatment		Shoot Zone Treatment		% of Corn Plants* with Stem-Twisting Injury	
EPTC lb/A	R-25788 lb/A	EPTC lb/A	R-25788 lb/A	Weeks after Treatment	
				2	4
Corn planted 2.5 cm deep and shoots exposed to 2.5 cm of treated soil.					
0	0	0	0	0	0
12	0	0	0	91	100
12	0	0	2	0	10
12	2	0	0	0	0
0	0	12	0	55	55
0	2	12	0	0	0
0	0	12	2	0	17
Corn planted 5 cm deep and shoots exposed to 5 cm of treated soil.					
0	0	0	0	0	0
6	0	0	0	27	45
6	0	0	1	0	0
6	1	0	0	0	0
0	0	6	0	100	100
0	1	6	0	86	86
0	0	6	1	0	0

* 12 DeKalb XL-44 corn plants were used for each treatment.

The results summarized in Table 3 show that application of EPTC only to the roots at 12 or 6 lb/A severely injured the corn, and good protection from this injury resulted when the antidote R-25788 was applied to the roots or to the shoots. In this test, the ratio of antidote to herbicide was 1 to 6, which is double the amount of antidote recommended and normally used. When the corn shoots were exposed to EPTC at 6 lb/A in 5 cm of treated soil, much more injury occurred than when exposed to EPTC at 12 lb/A in 2.5 cm of treated soil. This agrees with other studies which showed that more injury resulted in corn and other grasses from deeper planting and more exposure of the underground parts of the shoots to EPTC-treated soil. When the corn was planted 5 cm deep and EPTC at 6 lb/A was applied to the shoot zone, the most injury resulted. Appropriate placements of the antidote gave complete protection from injury by EPTC at 6 lb/A.

These results indicate that in deep planted corn where shoot exposure can cause severe injury, it is best for the antidote to be present in the same soil zone as the herbicide rather than be leached to lower depths around the roots.

In a further experiment, R-25788 was applied to the root or shoot zone of corn seedlings treated in the same or opposite zone with EPTC. No charcoal barriers were used. Circular plastic pots 15 cm in diameter and 17.5 cm deep filled to a depth of 16 cm with Scotts Valley loamy sand were used. For the 2.5 cm depth of planting, EPTC at 6 lb/A was incorporated with a small cement mixer into the upper 2.5 cm of soil in half of the pots for shoot exposure and in the lower 13 cm of soil in the other pots for root exposure. Twelve DeKalb XL-44A corn seeds were planted 2.5 cm deep in one set of pots and twelve DeKalb XL-45A corn seeds were planted in another set. For the 5 cm depth of planting, EPTC at 6 lb/A was incorporated into the upper 5 cm of soil in some of the pots and in the lower 10.5 cm of soil in the other pots. The antidote R-25788 was incorporated with the EPTC in the same zone in some pots and incorporated into the opposite zone in others. The experiment was repeated two more times and the results are reported in Table 4.

The results in Table 4 show that in three different tests where EPTC at 6 lb/A was applied to the shoot zone in the upper 2.5 or 5 cm of soil, all of the DeKalb XL-44A and XL-45A corn plants were injured severely, and application of R-25788 at 0.5 lb/A to the root zone gave partial protection from this injury. R-25788 gave complete protection under all other placement conditions including those where R-25788 was applied together with the EPTC in the shoot zone or in the root zone and where the antidote was applied to the shoot zone when EPTC was applied to the root zone.

The results indicate that R-25788 will offer good protection to these susceptible corn hybrids as long as it remains in the same soil zone as the EPTC.

TABLE 4

Effect of R-25788 when applied to the soil in the root or shoot zone of corn plants treated in the same or opposite zone with EPTC

Root Zone Treatment		Shoot Zone Treatment		% of Corn Plants* with Stem-Twisting Injury 4 Weeks after Treatment		
EPTC lb/A	R-25788 lb/A	EPTC lb/A	R-25788 lb/A	Test 1**	Test 2	Test 3
DeKalb XL-45A corn planted 2.5 cm deep and shoots exposed to 2.5 cm of treated soil.						
0	0	0	0	0	0	0
0	0	6	0	100	100	100
0	0.5	6	0	42	100	25
0	0	6	0.5	0	0	0
6	0	0	0			83
6	0.5	0	0			0
6	0	0	0.5			0
DeKalb XL-45A corn planted 5 cm deep and shoots exposed to 5 cm of treated soil.						
0	0	0	0	0	0	0
0	0	6	0	100	100	100
0	0.5	6	0	60	80	58
0	0	6	0.5			0
6	0	0	0			92
6	0.5	0	0			0
6	0	0	0.5			0

*   12 plants were used for each treatment.
** DeKalb XL-44A corn was used in Test 1.

IV. APPLICATION BY THE SPLIT-ROOT TECHNIQUE

An experiment was carried out to determine if application
of the antidote R-25788 to one root of a corn plant would
offer protection when EPTC was applied to another root. A
twin-pot was prepared by taping together two circular plastic
pots which were 15 cm in diameter and 17 cm deep. Both halves
of the twin-pot were filled to within 1 cm of the top with
loamy sand. A 7-day-old corn seedling (Figure 1) was transplanted to a twin-pot so that the primary root was placed in
one side (Tub A) and the two adventious roots were inserted
in the other side (Tub B) as shown in Figure 2.

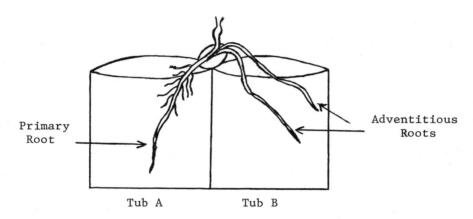

*Fig. 2. Split-root technique in which two plastic tubs
were taped together to form a twin-pot with the primary root
of a transplanted corn plant in the soil in Tub A and the two
adventitious roots in Tub B.*

A second plant was transplanted in the reverse position
with the two adventitious roots in Tub A and the primary root
in Tub B. A third plant was placed in the same position as
the first. Three more corn seedlings were transplanted to
another twin-pot with the adventitious roots of the two outside plants in Tub A and the primary root of the middle plant
also in Tub A. Two additional twin-pots were prepared as
replicates. Immediately after transplanting, Tub A of each
of four twin-pots was treated with a drench of EPTC at 6 lb/A
in 300 ml of Hoagland's nutrient solution. In two of these
twin-pots, the opposite Tub B was treated with 300 ml of
nutrient solution, and Tub B of the other two twin-pots received 300 ml of nutrient solution containing R-25788 at
1 lb/A.

Another experiment was carried out the same as described above except that the rate of EPTC was 12 lb/A and the rate of the antidote was 2 lb/A. R-25788 was also applied alone at 2 lb/A to Tub A and to Tub B in two additional twin-pots. The results of the two experiments are shown in Table 5.

TABLE 5

Effect of R-25788 in protecting corn from injury by EPTC when the two compounds were applied to different roots of the same plant

Primary Root Treatment		Adventitious Roots Treatment		% of Plants* with Leaf-Roll Injury Symptoms	
EPTC lb/A	R-25788 lb/A	EPTC lb/A	R-25788 lb/A	Weeks after Treatment	
				2	4
6	0	0	0	100	100
6	0	0	1	0	66
0	0	6	0	33	66
0	1	6	0	0	0
12	0	0	0	100	100
12	0	0	2	0	0
0	0	12	0	100	100
0	2	12	0	0	0

*   Three plants were used for each treatment.

The antidote alone at 2 lb/A had no effect so the results are not included in Table 5. The results in Table 5 show that treating one primary root or two adventitious roots of a corn plant with EPTC at 6 or 12 lb/A caused severe injury symptoms to develop. Treating the primary root with EPTC at 6 lb/A injured the plants more than treating the two adventitious roots. The corn was protected for two weeks or longer where the antidote R-25788 was applied at 1 and 2 lb/A to only one primary root or two adventitious roots when EPTC was applied to one or two other roots of the same corn plant at 6 and 12 lb/A, respectively.

These results show that the antidote does not have to be taken up by the same root as the herbicide in order to provide protection. This finding offers additional evidence that the mode of action of the antidote is not the inhibition of uptake of the herbicide by the corn roots.

V. APPLICATION AS FOLIAGE SPRAYS AND DIPS

This test was carried out to determine if the antidote
would protect corn seedlings from injury by EPTC when
applied as aqueous sprays to the foliage, as drops in the leaf
whorl, as a solution to a filter paper wrapped around the stem
at a site 0-10 mm above the coleoptilar node, and as a lanolin
paste around the stem at a site 0-10 mm above the coleoptilar
node. For these tests, 12 DeKalb XL-44A corn seeds were
planted in each of ten plastic pots as described earlier.
Eight days after planting the upper 2.5 cm of soil was removed
and replaced with a 7 mm layer of charcoal in 9 pots, and with
a 2.5 cm layer of charcoal in one pot. This left the coleoptilar node uncovered and unprotected in all pots except the
one with 2.5 cm of charcoal. A solution of 150 ml of EPTC
at 6 lb/A was applied to the root zone below the charcoal in
each of nine pots. One pot of plants was left untreated as a
control. A second pot received only EPTC in the roots. A
7 mm layer of vermiculite was placed over the charcoal of 4
other pots while they were sprayed with solutions of the antidote at 3.5, 10, and 25 lb/A at concentrations of 1, 4 and
10 mg/ml, respectively. Then the vermiculite was removed.
The seventh pot received 2 drops of a 10 mg/ml solution of
R-25788 in the whorl of the corn leaves at a rate of 10 lb/A.
In the eighth pot, the second leaf of each corn seedling was
dipped in a solution of 0.5 mg/ml of R-25788 contained in a
small vial supported on a stick. The underground parts of
the stems at the site 0-10 mm above the coleoptilar node were
wrapped with filter paper in the ninth pot. Then 0.2 ml of
a 10 mg/ml solution of R-25788 in water was applied to the
paper on two consecutive days at a rate of 24 lb/A. Each
plant in the tenth pot was treated in the same stem zone
(0-10 mm above the coleoptilar node) with approximately 2.5 mg
of R-25788 in 50 mg of lanolin paste at a rate of 15 lb/A.
The results are summarized in Table 6.

The results in Table 6 show that all methods of application of the antidote to different parts of the corn shoots
offered some protection to corn from injury by EPTC applied
to the roots. A foliage spray of R-25788 at 25 lb/A where
the underground parts of the stem were exposed protected all
of the plants for two weeks but after four weeks most of the
plants developed injury symptoms. The same spray to the
aboveground shoots only, where the underground parts were protected with 2.5 cm of charcoal, gave less protection and only
delayed injury for about one week. Drops of the antidote in
the whorl protected all of the plants for 2 weeks but the protection did not last. Leaf dips delayed symptoms for about
one week but did not offer much protection after 3 weeks.

Application on moist filter paper to the meristematic region
of the stem 0-10 mm above the coleoptilar node gave protection
for 2 weeks but the protection did not last. Application of
R-25788 to the same zone in lanolin paste gave the best and
longest protection of any treatment.

TABLE 6

*Effectiveness of R-25788 when applied as foliage
sprays, leaf dips, droplets in the whorl, drops on
filter paper, and in lanolin paste on the protection of corn from injury by·EPTC applied to the
roots at 6 lb/A*

EPTC Applied to Roots lb/A	Method of Applying R-25788 to Shoots	R-25788 lb/A	% of Corn Plants* with Leaf-Roll Injury Symptoms Weeks after Treatment		
			2	3	4
0	none	0	0	0	0
6	none	0	67	92	100
6	foliage spray**	3.5	33	83	100
6	" " **	10	18	63	82
6	" " **	25	0	42	83
6	" " ***	25	17	83	100
6	drops in whorl	10	0	22	78
6	leaf dip	<35	18	72	91
6	on filter paper	24	0	41	92
6	lanolin paste	15	0	0	20

  * 12 plants were used for each treatment.
 ** Sprayed aboveground and underground parts of shoot.
*** Sprayed only aboveground parts of shoot.

The long protection from the lanolin paste was attributed to continued absorption of the antidote by the stem for
several days from the paste. The R-25788-treated filter paper
dried out after several hours and absorption of the antidote
by the stem probably stopped. Previous injection tests indicated that injury from EPTC occurs only when the meristematic
region of the stem received a supply of the herbicide for
several days. The tests just described with the antidote
indicate that protection by the antidote also occurs only when
the same meristematic region is supplied with small amounts
of the antidote for several days.

VI. APPLICATION IN LANOLIN PASTE

     In the first of these experiments, EPTC in lanolin paste
was applied alone to different parts of the corn stems and
leaves to determine the site where the maximum injury occurs
which should be the site of action of the herbicide. The
antidote R-25788 in lanolin paste was also applied to different sites on the stems and leaves when EPTC was applied
to the roots to determine the site where the antidote exerts
its maximum protection.

     Ten corn seeds of each of three DeKalb hybrids (XL-43,
XL-45A and XL-64) were planted 2.5 cm deep in an upright position in each aluminum flat (15x22x7 cm) of soil as described
previously. Six days after planting, the upper 2.5 cm of soil
was removed and the soil was washed off the seedlings with a
syringe. A lanolin paste containing 2% EPTC was prepared by
dissolving 100 mg of technical EPTC in 4.9 grams of melted
lanolin and the mixture was stirred until cool. The paste
was applied with a small paint brush as a band 5 mm wide
around the underground parts of the corn stems at points 1-6
mm below the coleoptilar node, 0-5 mm above the node, and
5-10 mm above the same node. It was also applied to the
aboveground parts of the stem 15-20 mm above the coleoptilar
node and 25-30 mm above the node, as well as across the middle
of the first leaf, the upper half of the first leaf, and the
upper half of the second leaf. Both surfaces of the leaf were
painted.

     The above experiment was repeated with DeKalb XL-43 corn
except that a higher concentration of 4% EPTC in lanolin
paste was used and a 7 mm layer of charcoal slurry was placed
over the soil to prevent any contamination of the roots by
water of guttation passing over the lanolin. The flats were
watered by subirrigation to avoid contact of the lanolin with
water. The number of plants per treatment was increased to
30. The results of the two experiments are shown in Table 7.

     The results in Table 7 indicate that the site of action
of EPTC is the meristematic region of the stem located 0-5 mm
above the coleoptilar node, since this is the site where EPTC
in lanolin paste produced the most injury. This confirms the
results of the injection tests which also showed that the
maximum injury occurred when EPTC was applied to the meristematic region of the stems of the corn seedlings.

TABLE 7

Effect of applying EPTC in lanolin paste to different parts of corn seedlings to determine the site of action

Site Treated with EPTC in Lanolin Paste	% of Corn Plants Injured 5 Weeks after Treatment			
	2% EPTC Test 1*			4% EPTC Test 2**
	XL-43	XL-45A	XL-64	XL-43
None (control)	0	0	0	0
25-30 mm above coleop. node	0	0	0	0
15-20 mm " " "	0	0	0	7
5-10 mm " " "	0	0	0	33
0-5 mm " " "	70	10	50	100
1-6 mm below coleop. node	0	0	0	87
Middle of 1st leaf	0	0	0	
Upper half of 1st leaf	0	0	0	
Upper half of 2nd leaf	0	0	0	

*   10 plants were used for each treatment and no charcoal layer was used.
**   30 plants were used for each treatment and a charcoal layer protected the roots.

Two experiments were carried out using the antidote R-25788 in lanolin paste to determine the site of action. The experiments were carried out using the same methods used for the two previous experiments except that in each test, ten flats of corn seedlings were treated with a soil drench of EPTC at 6 lb/A in 400 ml of water applied to the roots. Then R-25788 in lanolin paste was applied to the same sites listed above for EPTC. The lanolin paste containing 5% R-25788 was prepared by dissolving 250 mg of technical R-25788 in 4.75 grams of melted lanolin with stirring. The treated flats were placed in larger aluminum pans for subirrigation. The heights of the plants were observed at weekly intervals and the percent stunting (inhibition of growth in height) is shown in Table 8.

The results in Table 8 show that when EPTC at 6 lb/A was applied to the roots of corn seedlings, severe stunting of the plants resulted. When EPTC was applied to the roots, application of 5% R-25788 in lanolin paste as a 5 mm band around the underground part of the stem located at the site 0-5 mm above the coleoptilar node gave the best protection of any site treated. This indicates that the site of action of the antidote is the same as that of the herbicide in the

meristematic region of the shoots of the corn seedlings located 0-5 mm above the coleoptilar node.

TABLE 8

*Effect of applying R-25788 in lanolin paste to dif-
ferent parts of corn seedlings after treating the
roots with EPTC at 6 lb/A*

Root Treatment EPTC lb/A	Site Treated with 5% R-25788 in Lanolin Paste	% Stunting of Corn Plants Test 1*	Test 2**
0	0	0	0
6	0	35	83
6	25-30 mm above coleop. node	10	57
6	15-20 mm " " "	10	24
6	5-10 mm " " "	5	8
6	0-5 mm " " "	0	0
6	1-6 mm below coleop. node	20	40
6	middle of 1st leaf	20	30
6	upper half of 1st leaf	20	35
6	upper half of 2nd leaf	20	75

 * Average for 10 plants of each of three corn hybrids
   (DeKalb XL-43, XL-45A and XL-64) 9 days after treatment
   with no charcoal layer.
** Average of 30 DeKalb XL-43 corn plants for each treatment
   measured 35 days after treatment using a charcoal layer
   to protect the roots.

Application of the antidote R-25788 to most parts of
leaves and stems of the corn seedlings gave some protection
from injury by the herbicide, indicating that some translocation of R-25788 upward and downward in the shoot to the
meristematic region took place. The next best protection
occurred when the antidote was applied 5-10 mm above the node
and this protection was much better than when applied 1-6 mm
below the node. This indicates that the antidote moves downward in the stem better than upward. Application of the antidote to the middle or upper half of the first leaf offered
considerable protection which was much better than application
to the upper half of the second leaf. This indicates that
translocation of R-25788 downward from the leaves also occured
when the antidote was applied to the leaf closest to the
meristems.

In Tests 1 and 2 of Table 8, the only plants that showed typical stem-twisting injury (not shown in Table 8) were those treated on the roots with EPTC alone without any antidote and those which received R-25788 in lanolin on the upper half of the second leaf. Because most lanolin applications of R-25788 protected the plants from this stem-twisting injury, the lesser injury symptom of stunting shown in Table 8 had to be used to determine the site where R-25788 offered the best protection.

VII. REFERENCES

1. R. A. Gray and A. J. Weierich, Weed Sci. 17, 223 (1969).
2. A. P. Appleby, W. R. Furtick and S. C. Fang, Weed Res. 5, 115 (1965).
3. J. H. Dawson, Weeds 11, 60 (1963).
4. E. L. Knake, A. P. Appleby and W. R. Furtick, Weeds 15, 228 (1967).
5. L. R. Oliver, G. N. Prendeville and M. M. Schreiber, Weed Sci. 16, 534 (1968).
6. C. Parker, Weeds 14, 117 (1966).
7. G. N. Prendeville, Weed Res. 8, 106 (1968).
8. L. Hoffmann, Weeds 10, 322 (1962).
9. L. Hoffmann, Abstracts Meeting Weed Sci. Soc. of Amer. 12 (1969).
10. O. C. Burnside, G. A. Wicks and C. R. Fenster, Weed Sci. 19, 565 (1971).
11. F. M. Pallos, R. A. Gray, D. R. Arneklev and M. E. Brokke, J. Agr. Food Chem. 23, 821 (1975).
12. F. Y. Chang, R. Stephenson and J. D. Bandeen, J. Agr. Food Chem. 22, 245 (1974).

PHYSIOLOGICAL RESPONSE OF LIPID COMPONENTS TO THIOCARBAMATES AND ANTIDOTES

Robert E. Wilkinson
The University of Georgia Agricultural Experiment Stations
Georgia Station, Experiment, Georgia

Thiocarbamate herbicide symptomology is consistent with a chronic loss of vital metabolic components on increasing toxicant concentration but not with acute blockage of a central reaction. S-Ethyl dipropylthiocarbamate (EPTC) activity is representative of the thiocarbamate herbicides. EPTC inhibits epicuticular wax synthesis, chloroplast fatty acid synthesis (FAS), "aged" tissue FAS and olelyl desaturase (DE) activity. These EPTC inhibitions of FAS and DE are reversed by 1,8-naphthalic anhydride (N) and N,N-diallyl-2,2-dichloroacetamide (R). Protein synthesis is not involved. EPTC-induced betacyanin efflux from "aged" red beet discs is time, temperature and concentration dependent and is reversed by N and R. Wheat root phospholipid content is decreased by EPTC and soybean leaf complex lipid class fatty acid contents are quantitatively and qualitatively altered by μM EPTC. The major response is a drastic decrease in linolenic acid (18:3) content. At high light intensities, wheat and corn growth is inhibited by EPTC. In darkness, neither crop is susceptible to EPTC at the same concentrations which induce growth reductions in the light. Differential penetration is not involved. EPTC inhibits wheat growth more than chlorophyll and carotenoid synthesis. Variation in wheat chloroplast electron carrier ratios indicates inhibition of epoxidation. Inhibition of polyphenol oxidase by EPTC is reversed by N and R. Thus, only the FAS portion of the total plant lipid synthesis is inhibited by the thiocarbamate herbicides. Mitochondrial phospholipid requirements for respiratory control and oxidative phosphorylation indicate that FAS inhibition is sufficient to explain thiocarbamate herbicide toxicity. However, the light intensity influence on plant responses to EPTC indicates the existence of more than one thiocarbamate herbicide mode of activity.

I.   INTRODUCTION

The various thiocarbamate herbicides are utilized in
many crops under highly diverse soil and environmental conditions. Although several hypotheses have been formulated,
thiocarbamate herbicide mode of action is unknown. Consequently, the mechanism(s) by which antidotes reverse the
toxicity of the various thiocarbamate herbicides is uncertain.
Because of the economic impact of these herbicides and antidotes, understanding their mechanism of activity is important
to agriculture throughout the entire North American continent.
Additionally, understanding the mechanism of thiocarbamate
herbicide activity may permit research to attack specific
sites of activity with more efficacious materials having
shorter half-lives and reduced environmental impact.

Thiocarbamate herbicide symptomology is diverse. In
monocots a major response is failure of the primary leaves to
penetrate the coleoptile or unroll if they do emerge through
the coleoptile. In dicots the response is a generalized
growth reduction. Both classes of plants respond in a normal
dose-effect pattern and in each case response to the thiocarbamate herbicide is not immediate but develops while the
plant is exposed to the herbicide. This pattern is indicative
of a phytotoxicity that is not incurred at the site(s) of
action but develops as a result of the gradual retardation
of an essential reaction in a separate system.

Such a generalized growth retardation has been demonstrated in yeast (Saccharomyces cerevisae Meyen) mutants
which do not synthesize polyunsaturated fatty acids (PUFA).
In these mutants, growth is dependent upon the addition of
PUFA to the media. The PUFA are incorporated into phospholipids (PL) which are, in turn, incorporated into the mitochondria; in addition, mitochondrial oxidative phosphorylation is correlated with the quantity and quality of PUFA
present in the PL (Figure 1) (1-3).

In a similar reaction scheme, a reduction in the
production and/or utilization of PUFA in plant membranes
could induce the gradual growth retardation seen in plants
treated with thiocarbamate herbicides. What data substantiate
this hypothesis?

II.   EPICUTICULAR WAX SYNTHESIS

Gentner (4) noted that S-ethyl dipropylthiocarbamate
(EPTC) inhibits epicuticular wax synthesis in cabbage
(Brassica oleraceae L.) leaves in the bud at the time of

exposure. Still et al. (5) find that S-(2,3-dichloroallyl)
diisopropylthiocarbamate (diallate) and EPTC inhibit epicuticular wax in peas (Pisum sativum L.). In the same year,
we demonstrated similar responses in sicklepod (Cassia
obtusifolia L.) with EPTC (6). Table 1 illustrates the
physical decrease in Sudan IV staining lipids in sicklepod
leaflets after postemergence spraying with EPTC. Thus, a
major influence of thiocarbamates on external lipid synthesis
exists;but, what is the effect of the thiocarbamate herbicides on the internal plant lipids?

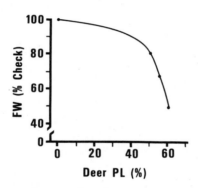

*Fig. 1. Decreased growth of Saccharomyces cerevisae
Meyen with decreasing phospholipid content.*

TABLE 1

*Sudan IV Staining Sicklepod (Cassia obtusifolia L.)
Leaflet Cuticle Thickness after EPTC Treatment (6)*

EPTC	Cuticle (μ)	
(kg/ha)	Upper	Lower
0	3.42ab[1]/	2.40a
0.125	3.55a	2.25ab
0.250	3.25abc	2.10bc
0.500	3.30ab	2.16abc
1.000	3.07abc	2.00bcd
2.000	2.90c	1.93cd
4.000	2.94bc	1.90d

[1]/Values in a column followed by the same letter are not
significantly different at the 5% level. Each value is the
average of 20 determinations.

III. INTERNAL FATTY ACID SYNTHESIS

A.   Isolated Chloroplasts

EPTC and diallate inhibit fatty acid synthesis (FAS) in isolated spinach (Spinacia oleracea L.) chloroplasts (Table 2) (7). Additionally, EPTC and diallate inhibit the incorporation of oleate-1-^{14}C but not palmitate-1-^{14}C into the oxygen esters of spinach chloroplasts (Table 3) (7).

TABLE 2

EPTC and Diallate Inhibition of Fatty Acid Synthesis in Isolated Spinach (Spinacia oleracea L.) Chloroplasts (7)

Herbicide	Fatty Acid Synthesis (%)	
(μM)	EPTC	Diallate
0	100	100
3	93	133
30	40	-
90	-	65
300	40	-

TABLE 3

Incorporation of Palmitate-1-^{14}C and Oleate-1-^{14}C into the Thio- and Oxygen-Ester Linkages by Spinach Chloroplasts (7)

Substrate	Herbicide	Conc	Esters	
		(μM)	Thio-	Oxygen-
		(%)	(%)	(%)
palmitate	0		100a[1]	100a
	EPTC	330	36b	113a
	diallate	90	63b	91a
oleate	0		100a	100a
	EPTC	330	36b	58b
	diallate	90	96a	54b

[1] Values in a column followed by the same letter are not significantly different at the 5% level.

Thus, EPTC inhibits the synthesis of fatty acids that
are inside the leaf. But the thiocarbamate concentrations
required to induce these fatty acid inhibitions are greater
than would be present in a plant after field exposure to EPTC
or diallate. This, again, appears to be an example of the
time-exposure concentration continuum response of plants to
the thiocarbamate herbicides. High concentrations for short
exposure periods often give the same general responses as
low concentrations for long periods of exposure.

Chloroplasts have a unique lipid complement that consists
of galactolipids plus phospholipids and, in chloroplasts, the
lipids show a very high linolenic acid (18:3) (i.e., a C_{18}
fatty acid with three double bonds at exact locations) content. The mechanism of linolenic acid synthesis in leaves
is still unknown. Only in a very few instances has linolenic
acid synthesis been attained in subcellular systems and this
does not include isolated chloroplasts. Therefore, isolated
chloroplasts are slightly suspect as a system in which to
demonstrate FAS inhibition by the thiocarbamate. Fortunately,
a non-chloroplast FAS system is available.

B.   "Aged" Tissues

The inhibition of plant FAS was corroborated and extended
to a nonchlorophyllous tissue when EPTC was shown to inhibit
the production of fatty acids in "aged" red beet (Beta
vulgaris L.) root discs (Table 4) (8). More interestingly,
1,8-naphthalic anhydride (N) increases production of linoleic
acid and N,N-diallyl-2,2-dichloroacetamide (R-25788 or R)
increases the production of oleic acid (Table 4). In these
tissues, synthesis of linoleic acid follows the sequence:
stearate→oleate→linoleate. In "aged" parenchyma tissue,
  (18:0)   (18:1) (18:2)
respiration, protein synthesis, and enzyme metabolism increase
(9). Included in these stimulated metabolic and protein synthesis patterns is an increase in oleyl desaturase activity
(10). Since the oleyl desaturase of "aged" parenchymatous
tissue is developed after tissue slicing and is associated
with protein synthesis (10), the increased linoleic acid
synthesis induced by N cannot be due to an inhibition of
protein synthesis. Moreover, the influence of R does not
entirely fit a protein synthesis pattern either, since
saturated and monoenoic FAS is present in "aged" sections
prior to slicing. R may limit olelyl desaturase activity but
it induces an increase in acid synthesis. Thus, three major
factors are demonstrated in these experiments; 1) EPTC
inhibits total FAS, 2) EPTC inhibits the desaturation of oleic

TABLE 4

*Influence of EPTC and Safeners on the Formation of
Fatty Acids in "Aged" Red Beet Discs (8).*

Chemical	(PPMW)	Total (%)	S[1]	Δ% M	D
-	0	100	-	-	-
E	37.8	77.6	+ 6.1a[2]	+ 4.7ab	-10.8bc
N	7.7	110.4	-20.9c	-11.7c	+32.6a
E+N	37.8+7.7	79.3	+0.2ab	- 4.2bc	+ 4.0b
R	7.7	107.6	- 5.5b	+ 1.4ab	+ 4.2b
E+R	37.8+7.7	90.3	+ 1.0ab	+11.0a	-12.0c
$S_{\bar{x}}$		4.6	3.26	3.31	4.87

[1] S = saturated, M = monoenoic, D = dienoic.

[2] Values in a column followed by the same letter are not
significantly different at the 5% level.

acid (18:1) to linoleic acid (18:2) by olelyl desaturase,
and 3) these general activities are reversed by the antidotes N and R.   If red beet root tissue mitochondria show
the same response to decreased PUFA in retarded oxidative
phosphorylation as was demonstrated for the yeast mutants,
then a means of explaining thiocarbamate toxicity is available.   This general hypothesis has been partially corroborated
by Bolton and Harwood (11) who demonstrated total FAS inhibition in "aged" potato (Solanum tuberosum L.) discs by μM
concentrations of EPTC, diallate, and S-(2,3,3-trichloroallyl)
diisopropylthiocarbamate (triallate).   However, Bolton and
Harwood (11) added a protein synthesis inhibitor (i.e.,
chloramphenicol) to their discs prior to "aging".   Thus, no
olelyl desaturase was synthesized, and the linoleic acid
content in all of their tissues was abnormally low.

Betacyanin efflux is concomitant with the influence of
EPTC on FAS inhibition in "aged" red beet root discs.   This
betacyanin efflux is time, concentration (Table 5), temperature (Table 6), and energy dependent (8).   Additionally,
betacyanin efflux induced by EPTC is reversed by N and R
(Figures 2 and 3).   Thus, an energy-dependent physiological
condition (i.e., betacyanin retention within the vacuole) is

TABLE 5

*Betacyanin Efflux (△OD) from Red Beet (Beta vulgaris L.)
Root Discs Exposed to Varying EPTC Concentrations (8).*

Time	EPTC (mM)			
(Hr)	0	0.19	0.95	1.9
	---------------- (△OD) ----------------			
0	0	0	0	0
1	0.01e[1]/	0.01e	0.01e	0.3d
2	0.01e	0.01e	0.01e	0.5c
3	0.01e	0.01e	0.01e	0.6bc
4	0.01e	0.01e	0.01e	0.7b
8	0.01e	0.01e	0.01e	0.72b
16	0.03e	0.01e	0.09e	0.85a
24	0.05e	0.09e	0.30d	1.00a

[1]/ Values followed by the same letter are not significantly
different at the 5% level. Each value is the average of
10 replications with ten 7 mm x 1 mm discs/replication bathed
in 0.1 M phosphate buffer at 25°C.

TABLE 6

*Temperature Influence on Betacyanin Efflux from 1.0×10^{-3}M
EPTC Treated Red Beet (Beta vulgaris L.) Root Discs (8).*

	Temperature ($^{\circ}$C)[1]/					
	15		20		25	
	EPTC		EPTC		EPTC	
Hours	-	+	-	+	-	+
	---------------------- (△OD) ----------------------					
0	0.00	0.00	0.00	0.00	0.00	0.00
1	0.00	0.01	0.05	0.22	0.02	0.57
2	0.01	0.05	0.06	0.38	0.02	0.88
4	0.02	0.05	0.08	0.45	0.03	0.90
12	0.03	0.08	0.06	0.60	0.03	0.92
24	0.03	0.08	0.08	0.63	0.18	0.93

[1]/ Test conditions as described in Table 5.

modified by EPTC, this action of EPTC is reversed by the
safeners, and all of these activities are concomitant with
equivalent FAS patterns.

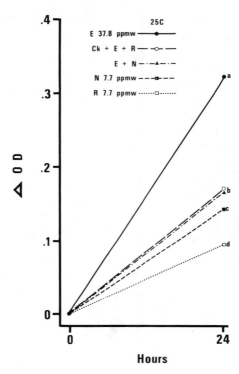

Fig. 2.   Influence of antidotes on betacyanin efflux from "aged" red beet (Beta vulgaris L.) root discs induced by EPTC.   Culture conditions same as listed in Table 5.   25°C. Points followed by the same letter are not significantly different at the 5% level.

The general concept of FAS inhibition as a major mechanism of action of μM concentrations of the thiocarbamate herbicides has been extended to most of the members of this herbicide family (7, 11, 12).   The antidotes N and R also inhibit FAS when applied at herbicidal rates (12).   Thus, harkening back to the yeast mutants whose growth was decreased when PUFA were decreased in the medium with a concomitant decrease in PUFA in mitochondrial PL (1-3), the question of the influence of EPTC on PL synthesis arose.

IV.   COMPLEX LIPID SYNTHESIS

Wheat (Triticum sativum L.) root PL content is decreased more than fresh weight production by μM EPTC concentrations

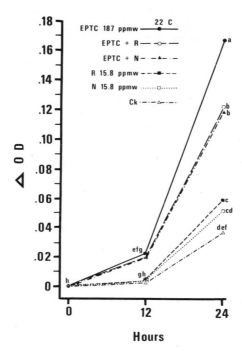

Fig. 3.   Influence of antidotes on betacyanin efflux from "aged" red beet (Beta vulgaris L.) root discs induced by EPTC.   Culture conditions same as those listed for Table 5.   22°C.   Points followed by the same letter are not significantly different at the 5% level.

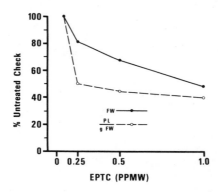

Fig. 4.   Influence of preemergence applied EPTC on phospholipid content and fresh weight in wheat (Triticum aestivum L.) roots (13).

(Figure 4) (13). This is associated with a general decrease
in FAS (60%) and a greater inhibition of linolenic acid synthesis (70%) (13). Soybean (<u>Glycine max</u> (L.) Merr.) leaf
fatty acid content decreases 64% while fresh weight decreases
18% (Table 7) (14). Associated with this decrease in total
fatty acid content is an 83% decrease in linolenic acid
content (Table 7) (14). This decrease in linolenic acid content is partially offset by increased oleic acid content and
unchanged linoleic acid content. Consequently, the major
influence noted is an inhibition of linolenic acid synthesis.
There is a decrease in all constituent lipid class totals
except PL (Table 8) (14). However, the PL group contained
phosphatidylglycerol, phosphatidylserine, phosphatidylethanol-
amine, phosphatidylcholine, and free fatty acids.

After further purification, phosphatidylcholine is shown
to increase almost three fold while the quality of the fatty
acid composition is greatly altered with decreases in PUFA
(Table 9). Since phosphatidylcholine is a precursor of

TABLE 9

Phosphatidylcholine Fatty Acids from Soybean
(Glycine max Merr.) Leaves after EPTC Treatment

EPTC (μM)	Palmitic	Stearic	Oleic	Linoleic	Linolenic	Total Fatty Acid μg/g FW
	-------------------------%--------------------					
0	33.1	9.3	9.0	24.7	27.9	685
2.6	20.8	4.3	28.2	37.4	9.3	1878

galactolipid synthesis in the chloroplast (15, 16), these data
suggest a major disruption of the lipid quality of chloroplast and mitochondrial membranes. Associated electron
transfer inhibitions should result. However, it now appears
that chloroplast lipids are synthesized inside the chloroplast and utilized therein. Non-chloroplast lipids are,
apparently, produced by a separate system. Further, the
"time of exposure vs concentration" continuum is demonstrated
again. Long exposure periods at low concentrations produce:
1) major total lipid content changes, and 2) major lipid
quality alterations.

Since lipids are requisite for the photo-oxidation of
water by chloroplasts (17, 18), as well as oxidative phosphorylation in mitochondria, this requirement for lipids in

TABLE 7

EPTC Influence on Soybean Leaf Total Fatty Acids.
Each value is the average of ten analyses (14).

EPTC	Fresh weight % untreated	% Fatty Acid Composition					Total	$\dfrac{Linolenic}{Oleic + linoleic}$	Σ U[1]/Σ S
		Palmitic	Stearic	Oleic	Linoleic	Linolenic			
		-----μg/g FW-----							
0	100	17.2	3.3	2.1	9.9	67.5	3757$\pm$100	5.62	3.87
2.6	82	23.5	4.7	16.5	23.8	31.5	1385$\pm$80	0.78	2.55

[1]/ΣU/ΣS = total unsaturated fatty acids / total saturated fatty acids

TABLE 8

Fatty Acid Content of Lipid Classes from Mature Soybean Leaves.
Each value is the average of ten analyses (14).

Lipid[2]	EPTC µM	Palmitic	Stearic	Oleic	Linoleic	Linolenic	Total (µg/g FW)	Linolenic / Oleic+Linoleic	ΣU[1] / ΣS	Class Total %
		%---								
MGDG	0	14.7	4.0	18.0	23.8	39.3	122	1.06	4.34	(22.6)
	2.6	23.7	6.6	14.5	17.1	38.2	76	0.83	2.30	(9.9)
CL	0	17.2	6.9	12.1	20.7	43.1	58	1.31	3.15	(11.2)
	2.6	30.3	12.1	6.1	15.2	36.4	33	1.71	1.36	(5.0)
DGDG	0	17.2	10.2	6.2	12.5	53.9	128	2.88	2.65	(25.1)
	2.6	20.7	7.4	12.4	24.8	34.7	121	0.93	2.56	(15.5)
SL	0	20.0	9.1	5.4	7.3	58.2	55	4.58	2.44	(10.5)
	2.6	18.8	7.7	26.0	36.1	11.5	208	0.18	2.78	(25.9)
PL	0	35.3	18.1	5.9	19.8	30.9	136	1.20	1.30	(30.5)
	2.6	23.5	5.2	25.9	33.4	11.9	344	0.20	2.48	(43.8)

[1] ΣU = total unsaturaged fatty acids, ΣS = total saturated fatty acids.
[2] MGDG = monogalactosyldiglycerides, CL = cardiolipids, DGDG = digalactosyldiglycerides, SL = sulfolipids, PL = phospholipids.

chloroplasts might function as a means for the physical
separation of different reactive centers on each side of a
chloroplast membrane with the carotenoids and xanthophylls
functioning as trans-membrane electron carriers (17, 18).
Coupled with this concept of the function of lipids in membranes is the report by Montague and Ray (19) that PL outside
the chloroplasts are produced in the Golgi bodies and endoplasmic reticulum. Since nonchloroplastic olelyl desaturase
is located in the microsomes (20, 21), the general pattern in
nonchloroplast systems is: a) synthesis of saturated fatty
acids in mitochondria, b) transfer of the fatty acids to
microsomes where they are desaturated, c) transfer of the unsaturated fatty acids to Golgi bodies and the endoplasmic reticulum where the PL are synthesized, and d) transfer of the
PL back to the mitochondria where the major portion of PL
containing linoleic acid accumulation in "aged" tissues
occurs (20, 21).

V.   LIGHT INTENSITY INFLUENCE

If inhibition of FAS is a major mode of activity for the
thiocarbamate herbicides, variation in the factors requisite
for PUFA synthesis should show an influence on thiocarbamate
activity. Chloroplast PUFA synthesis requires light as a
means of providing energy and oxygen. So, the influence of reduced light intensity was examined on the response of plants
to EPTC and the concomitant effect on FAS. Surprisingly, dark
grown wheat (Triticum aestivum L.) was found to be totally
unresponsive to EPTC concentrations which induce a 90% growth
inhibition (i.e., height) in "high" light intensities (i.e.,
about 20% full sunlight) (Table 10) (22). This growth
inhibition is also reflected in the fresh weight data (Table
11) (22). The plants in all three light intensities emerge
at the same time. The leaves of the plants treated with 125
ppbw EPTC and grown under "high" light intensities never
penetrate through the coleoptile. Thus, penetration-absorp-
tion phenomena are excluded in this response (22). Additionally, corn (Zea mays L.) exhibits a similar height and fresh
weight response pattern between EPTC and light intensity
(Figures 5 and 6). This response does not require chlorophyll since chlorophyllous and albino corn plants exhibit
the same percentage inhibition of growth by EPTC (Figure 7).
However, it appears that the response is not a simple light
intensity pattern. The wheat data were taken from growth-
chamber grown plants. After completion of this set of experiments, the fluorescent lamps in the growth chamber were
exchanged for a second type of lamps with a supposedly higher
light intensity but a different spectra. Subsequent tests
in this growth chamber showed no influence of light on the

TABLE 10

*Wheat (Triticum aestivum L. var. Holley) Height
after Preemergence Exposure to EPTC (22)*

EPTC (ppbw)	Dark	Height (% Untreaded Check) Low[1] Light Intensity	High Light Intensity
0	100a [2]	100a	100a
15.6	100a	90b	75c
31.25	100a	80c	65d
62.5	100a	62d	37f
125.0	100a	52e	8g

[1] Low light intensity = 20 μeinsteins/m^2 sec
High light intensity = 270 μeinsteins/m^2 sec

[2] Values followed by the same letter are not significantly different at the 5% level. Each value is the average of 5 replications using 25 seeds in 500 g sand into which EPTC had been incorporated.

TABLE 11

*Fresh Weight/Pot of Wheat (Triticum aestivum L. var.
Holley) Grown for 14 Days in Sand into Which EPTC
was Incorporated (22)*

EPTC (ppbw)	Fresh Weight/Pot (% Untreated Check) Dark	Light Intensity Low[1]	High
0	100a[2]	100a	100a
15.6	100a	77b	45c
31.25	100a	45c	22d
62.5	100a	20d	10de
125.0	100a	18d	0e

[1] Low light intensity = 20 μeinsteins/m^2/sec
High light intensity = 270 μeinsteins/m^2 sec

[2] Values followed by the same letter are not significantly different at the 5% level. Culture conditions same as Table 10.

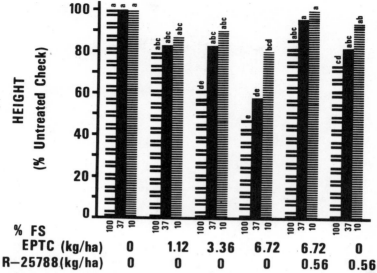

Fig. 5. *Height response of corn (Zea mays L.) after preemergence exposure to EPTC, R-25788, or mixtures after 14 days in the greenhouse. Points followed by the same letter are not significantly different at the 5% level.*

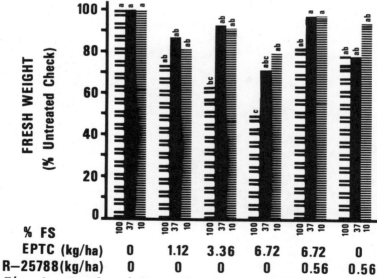

Fig. 6. *Fresh weight production of 14-day old corn (Zea mays L.) grown in the greenhouse after preemergence exposure to EPTC, R-25788, or mixtures. Points followed by the same letter are not significantly different at the 5% level.*

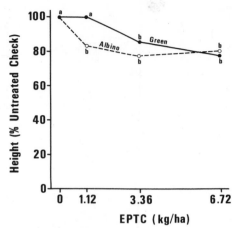

Fig. 7. Height of chlorophyllous and albino corn (Zea mays L.) after growth for 10 days in the greenhouse and pre-emergence exposure to EPTC. Points followed by the same letter are not significantly different at the 5% level.

response of wheat to EPTC. Indeed, wheat grown under the new lamps does not respond to EPTC. Thus, it appears that there is a discrete light quality phenomenon involved here also. Finally, there are major response differences between wheat and corn. Wheat shows a response at 20% of full sunlight intensity. Corn requires full summer light intensities to demonstrate this light intensity response. The muted response of corn to different light intensities after exposure to EPTC at other seasons of the year may be a further example of a light quality response. Much research is required.

VI. CHLOROPLAST CONSTITUENT SYNTHESIS

Chloroplasts contain many lipid constituents other than fatty acids and many of these other constituents are synthesized by systems which utilize the same precursors and co-enzymes as are utilized by fatty acid synthetase. Since the quantity of plant lipids present at any given time is a composite of synthesis and degradation and fatty acid synthesis and degradation both require energy (i.e., electrons), the total information available suggested that EPTC and the antidotes were influencing some portion of the electron transfer system that was totally overloaded in high light intensities. Consequently, analyses were undertaken of the chloroplast pigments and the various electron carrying quinones in EPTC-treated wheat grown under "high" light intensities (i.e., 2000 ft-can.).

Although growth is inhibited (Tables 10 and 11), chloro-phylls a and b contents (μg/g F.W.) increase as EPTC concen-tration is increased (Figure 8) (22). Concomitantly, the total carotenoid content is not greatly altered. These observations are explicable on the basis of growth inhibition by EPTC without an accompanying metabolic inhibition of chlorophyll production. This concept has been corroborated in germinating moss (Polytrichum commune L.) spores (23). Since the carotenoids are a highly complex group and three of the xanthophyll components have been shown to function in an electron carrying capacity (Figure 9) (24), an evaluation of the influence of EPTC on leaf violaxanthin cycle component content was undertaken. EPTC does not appreciably alter the content of the most oxidized member of the cycle (i.e., violaxanthin), but the quantities of the more reduced cycle participants are increased as EPTC is increased (Figure 10) (22). The rate of increase of the intermediately reduced antheraxanthin is less than that of the fully reduced zeaxanthin. Thus, these data suggest: 1) an inhibition by EPTC of epoxidation of zeaxanthin and antheraxanthin, 2) an inhibition of oxygen utilization, or 3) a reduced capacity of these highly lipophilic membrane-bound components to migrate through the membrane.

Other electron-carrying quinone systems are present in chloroplasts. Plastoquinone functions in the photosynthetic chain as an electron carrier between photosystems I and II. Analyses of plastoquinone reveal similar results with accumu-lation of plastohydroquinone (Figure 11) (25). Although the function of α-tocopherol has not been definitely established, it is postulated to function as an electron carrier (26). The contents of α-tocoquinone and α-tocopherol substantiate these altered electron-carrying component patterns in yet another electron-carrying system (Figure 12) (25). Consequently, these data demonstrate: 1) a lack of influence of EPTC on the syntheses of terpenoid constituents which utilize the same precursors and some of the same -SH containing enzymes re-quisite to FAS, and 2) a possible influence on the activity of various electron-carrying systems.

VII. MIXED-FUNCTION MONOOXYGENASE ACTIVITY

Fatty acid synthesis, fatty acid desaturase, and viol-axanthin cycle component interconversion require energy and oxygen. All of these systems exist in or on membranes. Therefore, it was of interest to measure the influence of EPTC on a nonmembrane-bound mixed-function monooxygenase. Polyphenol oxidase (PPO) utilizes oxygen and catechol to produce p-benzoquinone but is not membrane bound and does not

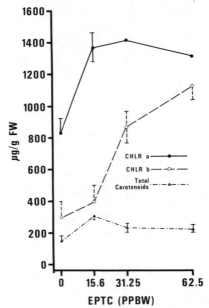

*Fig. 8.   Chlorophyll a, chlorophyll b, and total caro-
tenoid content of 14-day old wheat (Triticum aestivum L.)
leaves after preemergence exposure to EPTC.*

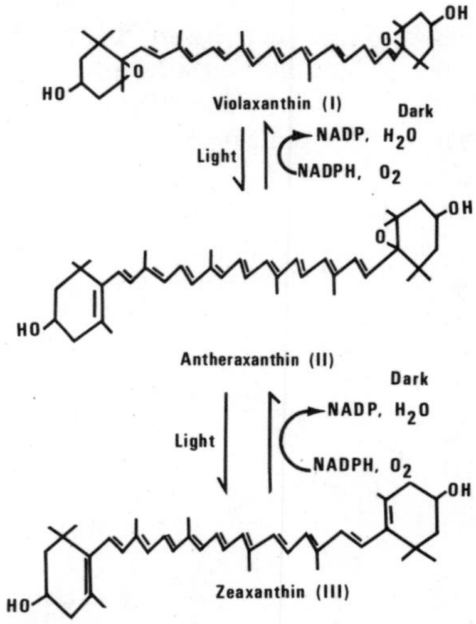

*Fig. 9.   Violaxanthin cycle components and interconver-
sion requirements (24).*

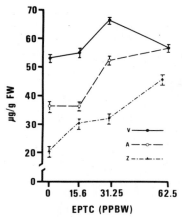

Fig. 10.   Influence of preemergence EPTC application on
the violaxanthin, antheraxanthin, and zeaxanthin content of
14-day old wheat (Triticum aestivum L.) leaves.

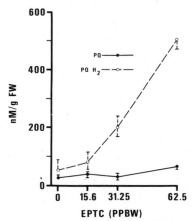

Fig. 11.   Plastoquinone and plastohydroquinone content
of 14-day old wheat (Triticum aestivum L.) leaves after preemergence exposure to EPTC.

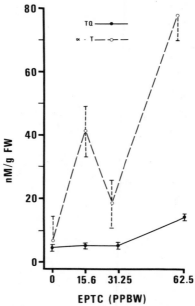

Fig. 12.   α-Tocoquinone and α-tocopherol contents
of 14-day old wheat (Triticum aestivum L.) leaves after
preemergence exposure to EPTC.

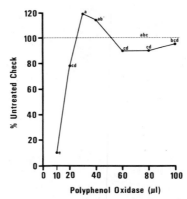

Fig. 13.   Polyphenol oxidase (PPO) activity with EPTC.
PPO = 1 activity unit/μl, catechol = 3 mM, proline = 60 mM,
0.2 mM phosphate buffer - pH 6.5; 22°C; EPTC = 1.9 x 10^{-3}M;
3.0 ml total volume. Points followed by the same letter are
not significantly different at the 5% level.

require an external energy source. EPTC inhibits the activity
of low PPO concentrations (Figure 13) (22). Antidotes N and
R reverse the inhibition of PPO activity by EPTC (Table 12).

TABLE 12

*Influence of EPTC, and Antidotes, and Mixtures
on the Activity of Polyphenol Oxidase*

PPO (µl)	Catechol M	Chemical		PPO Activity (%)
		(Name)	(PPMW)	
10	0.2	-	-	100
		E	187	22
		N	15.8	98
		E+N		103
		R	15.8	86
		E+R		133

However, EPTC inhibition of PPO activity was shown to be
competitive with catechol concentration (22). More than
a simple membrane phenomenon consisting of lipid component
solubilities and transferrals seems to be involved here. An
influence on oxygen utilization has not been eliminated.
Moreover, the requirement for membrane lipids in oxidative
phosphorylation or photosynthesis has not differentiated
between: 1) an active participation of the lipid molecules
in the reaction, and 2) a passive spacing phenomenon wherein
the lipids retain the active enzyme sites in a requisite
spatial configuration. Generally, a lack of lipid class
specificity in these lipid requirements has tended to support
a passive spatial role for the lipids in these systems, but
critical definition has not been attained. Consequently,
the activity of the thiocarbamate herbicides and antidotes on
the PPO system does not eliminate a membrane affect syndrome.
However, it does relegate the membrane affect to a secondary
response phenomenon.

VIII.   DISCUSSION

        Fatty acid, carotenoid, and quinone syntheses utilize
sulfhydryl containing enzymes and common precursors. The
inhibition of FAS without a concomitant inhibition of caroteno-
noid or quinone synthesis argues against a general sulfhydryl
enzyme deactivation as the major activity of the thiocarbamates. The stimulation of carotenoid and quinone syntheses associated with FAS inhibition may exhibit a simple feed-

back phenomenon. Mitochondrial PL requirements for respiratory control and oxidative phosphorylation indicate that
FAS inhibition is sufficient to explain thiocarbamate herbicidal activity.

However, the light intensity and light quality influences on plant responses to the thiocarbamate herbicides are
indicative of the existence of more than one mode of activity
for this class of compounds. In this connection, it must be
remembered that the gibberellin GA_3 reverses EPTC-induced
growth inhibitions in corn (27-29). Gibberellins have been
reported to alter PL synthesis and stimulate the activity of
enzymes which generate phosphatidylcholine, phosphatidyl-
ethanolamine, and phosphatidylinositol from phosphatidic acid
and phosphatidylglycerol (30). But, GA does not increase the
total PL present in the cells (30). Thus, the increased
PUFA content of "aged" red beet discs does not fit a gibberellin pattern. Contrarywise, "aged" artichoke (Helianthus
tuberosus L.) tuber tissue has been shown to synthesize
gibberellin (31) and the synthesis is   temperature dependent
with increased gibberellin contents at $4^{\circ}C$ as opposed to $20^{\circ}C$.
These data do fit the responses on betacyanin efflux (8).
Finally, Varty and Laidman (32) disagree with the reported
GA_3 altered PL synthesis (30, 31). They find an inhibition
rather than a stimulation of PL synthesis by GA_3 in wheat
aleurone tissue.

In view of the differential responses of growth and
metabolism, these gibberellin effects do not appear to
explain EPTC toxicity. But, the gibberellins influence
membranes, enzyme function, and protein synthesis (33).
These, in turn, may ultimately explain thiocarbamate toxicity.
Thus, the influence of thiocarbamate herbicides and antidotes on the growth response of plants to the gibberellins
in conjunction with the influence of PUFA and gibberellins on
membrane function remain a most challenging research project.

Finally, the concept of the conversion of a thiocarbamate
herbicide into a more active chemical molecule which in turn
is the major toxicant is a most attractive hypothesis. Selectivity, specificity, and activity may be enhanced thereby.
But the total experience in herbicide biochemical activity
shows that the mode of activity is the same for each member
of a herbicide family of compounds regardless of the
general sensitivity or resistance of field grown plants.
Resistance or sensitivity are usually explained by differential absorption or degradation patterns. Consequently, what-
ever the active chemical toxicant may be, a search will
continue for biochemically consistent reactivities extending

throughout the thiocarbamate herbicide family and across susceptible and resistant plant species categories. Because of the widely diverse physiological and biochemical responses that occur following thiocarbamate applications, some very general biochemical activity will be requisite if a common activity is to be visualized. Lipid influences on membrane activities are sufficiently general to fit this pattern.

IX. REFERENCES

1. Haslam, J. M., J. W. Proudlock, and A. W. Linnane, Bioenerg. 2, 351 (1971).
2. Haslam, J. M., T. W. Spithill, A. W. Linnane, and J. B. Chappel, Biochem. J. 134, 949 (1973).
3. Sangot, M., G. S. Cobon, J. M. Haslam, and A. W. Linnane, Arch. Biochem. Biophys. 169, 577 (1975).
4. Gentner, W. A., Weeds 14, 27 (1966).
5. Still, G. G., D. G. Davis, and G. L. Zander, Plant Physiol. 46, 307 (1970).
6. Wilkinson, R. E., and W. S. Hardcastle, Weed Sci. 18, 125 (1970).
7. Wilkinson, R. E., and A. E. Smith, Weed Sci. 23, 100 (1975).
8. Wilkinson, R. E., and A. E. Smith, Weed Sci. 24, 235 (1976).
9. Stevenick, R. F. M. van, Ann. Rev. Plant Physiol. 26, 237 (1975).
10. Willemot, C., and P. K. Stumpf, Can. J. Bot. 45, 579 (1967).
11. Bolton, P., and J. L. Harwood, Phytochem. 15, 1507 (1976).
12. Wilkinson, R. E., and A. E. Smith, Weed Sci. 23, 90 (1975).
13. Karunen, P., and R. E. Wilkinson, Physiol. Plant. 35, 228 (1975).
14. Wilkinson, R. E., B. Michel, and A. E. Smith, Plant Physiol. 60, 86 (1977).
15. Jayard, J., and R. Douce, Biochim. Biophys. Acta 486, 273 (1977).
16. Slack, C. R., P. G. Roughan, and N. Balasingham, Biochem. J. 162, 289 (1977).
17. Mangel, M., D. S. Berns, and A. Ilani, J. Membr. Biol. 20, 171 (1975).
18. Toyoshima, Y., M. Morino, H. Motoki, and M. Sukigara, Nature 265, 187 (1977).
19. Montague, M. J., and P. M. Ray, Plant Physiol. 59, 225 (1977).
20. Abdelkader, A. B., and P. Mazliak, Eur. J. Biochem. 15, 250 (1970).

21. Abdelkader, A. B., P. Mazliak, and A. M. Catesson, Phytochem. 8, 1121 (1969).
22. Wilkinson, R. E., Bot. Gaz. 138, in press (1977).
23. Karunen, P., N. Valanne, and R. E. Wilkinson, The Bryologist 79, 332 (1976).
24. Siefermann, D., and H. Y. Yamamoto, Arch. Biochem. Biophys. 171, 70 (1975).
25. Wilkinson, R. E., Pesticide Biochem. Physiol., in press (1977).
26. Barr, R., and F. L. Crane, Plant Physiol. 59, 433 (1977).
27. Harvey, B. M. R., F. Y. Chang, and R. A. Fletcher, Can. J. Bot. 53, 225 (1975).
28. Donald, W. W., and R. C. Fawcett, Proc. North Central Weed Control. Conf. 30, 28 (1975).
29. Donald, W. W., R. G. Harvey, and R. S. Fawcett, Proc. Weed Sci. Soc. Amer., 92 (1977).
30. Koehler, D., K. D. Johnson, J. E. Varner, and H. Kende, Planta 104, 267 (1970).
31. Bradshaw, M. J., and J. Edelman, J. Exp. Bot. 20, 87 (1969).
32. Varty, K., and D. L. Laidman, J. Exp. Bot. 27, 748 (1976).
33. Browning, G., and P. F. Saunders, Nature 265, 375 (1977).

Part V

Antidote Metabolism and Effects on Herbicide Metabolism

METABOLISM OF R-25788 (<u>N</u>,<u>N</u>-DIALLYL-2,2-
DICHLOROACETAMIDE) IN CORN PLANTS, RATS AND SOIL

J. Bart Miaullis, Victor M. Thomas, Reed A. Gray,
John J. Murphy and Robert M. Hollingworth
Stauffer Chemical Company

*<u>N</u>,<u>N</u>-Diallyl-2,2-dichloroacetamide (R-25788) is an
effective protectant for corn against injury by thiocarbamate
herbicides. Metabolism and environmental studies using the
labeled compound, <u>N</u>,<u>N</u>-diallyl-2,2-dichloro(2-^{14}C)acetamide,
have been done with corn plants grown in treated soil, with
orally dosed rats, and in two soil types. Four days after
a single oral dose, rat urine, feces, expired $^{14}CO_2$, and
tissue residues accounted for 73%, 13%, 5%, and 6%,
respectively, of the administered radiocarbon. Corn
seedlings liberated 6% of the absorbed radioactivity as
$^{14}CO_2$ in a ten day study. The major metabolites in both rats
and corn plants were formed by dechlorination and oxidation
of the antidote. R-25788 was identified in corn seedling
extracts and rat urine although in both cases it accounted
for minor amounts of the total radiocarbon. In soil, 26%
of the radioactivity was liberated as $^{14}CO_2$ in 21 days. The
major products remaining in the soil were R-25788 along
with at least 7 minor metabolites. Three of the non-polar
metabolites were identified as <u>N</u>-allyl-2,2-dichloroacetamide,
<u>N</u>,<u>N</u>-diallyl-2-chloroacetamide, and <u>N</u>,<u>N</u>-diallylacetamide.*

I. INTRODUCTION

The compound N,N-diallyl-2,2-dichloroacetamide (Stauffer
Chemical Co., R-25788) has been effective in decreasing the
susceptibility of corn plants to the thiocarbamate injury
(1). With the introduction of any new chemical comes concern
about the safety of the product and its environmental fate.
Metabolism studies using radiotracer techniques have become
an integral part of our attempt to answer these questions.
The metabolism studies reported here for R-25788 have been
carried out in rats, corn plants and soils to determine the
fate of R-25788 and perhaps provide some clues as to the
mode of action of the compound.

Metabolism of the herbicide CDAA, which is structurally
similar to R-25788 but having one less chlorine, has been
reported for plants (2). Those studies showed the chloro-
acetic acid liberated upon hydrolysis was metabolized to

glycolic acid by the corn plants. In the present studies, R-25788 was labeled with ^{14}C using a dichloro[2-^{14}C]acetyl chloride intermediate to give the final [^{14}C]R-25788 as shown in Figure 1.

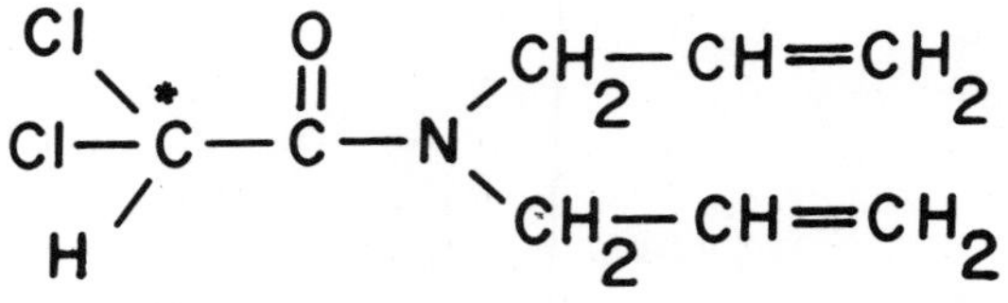

N,N— Diallyl—2,2 — dichloro—

[2-^{14}C] acetamide

Fig. 1. Structure and label position of [^{14}C]R-25788.

II. MATERIALS AND METHODS

A. Chemicals, Radiocarbon Assay and Thin-Layer
    Chromatography

Radiolabeled [^{14}C]R-25788 (N,N-diallyl-2,2-dichloro-[2-^{14}C]acetamide was synthesized from commercially available dichloro[2-^{14}C]acetyl chloride (Mallinckrodt,St. Louis, MO) and diallylamine by Jules Kalbfeld of the Stauffer Chemical Company, Western Research Center, Richmond, California. Radiopurity of the final material was 98% as determined by thin layer chromatography (tlc) in ethyl ether:hexane 1:1 (EH) and benzene:chloroform 1:1 (BC). The specific activity was 4.2 mCi/mMole.

Reference compounds and metabolites were synthesized or supplied by Stauffer Chemical Company, Western Research Center, except for the following: [^{14}C]glycolic acid and [1,2-^{14}C]oxalic acid were purchased from ICN (Cleveland, OH) and dichloro[2-^{14}C]acetic acid from Mallinckrodt (St. Louis, MO).

Radiocarbon determinations were made with a Model 3375 Packard Liquid Scintillation (lsc) Spectrometer (Packard Instrument Company, Downers Grove, IL). Liquid scintillation fluid was prepared with 3 liters toluene, 1 liter Triton® X-100 (Rohm and Haas, Philadelphia, PA), 16.5 g 2,5-diphenyl-

oxazole and 0.3 g 1,4-_bis_-2-(4-methyl-5-phenyloxazole)benzene.
For ^{14}C analysis, aliquots of liquid samples were dissolved
directly in 5 ml of lsc fluid while ^{14}C on silica gel scraped
from thin layer chromatography plates was suspended in the
fluid by adding water to form a gel before analysis. Solid
samples except for charcoal traps were combusted in a
Packard Model 305 sample oxidizer and the ^{14}C was determined
by lsc of the trapped [14]CO_2. Efficiency determinations for
[14]C analysis were done with [^{14}C]toluene internal standard-
ization for lsc samples and with [^{14}C]hexadecane for combusted
samples.

B.   Rat, Plant and Soil Metabolism

Simonsen albino Sprague-Dawley derived rats (Simonsen
Laboratories, Gilroy, CA) were dosed by oral intubation with
a 50% aqueous ethanol solution of [^{14}C]R-25788 and unlabeled
R-25788 (99.3%ai). Doses given the animals were determined
from the radiocarbon content of the dosing solutions. Final
specific activity of the [^{14}C]R-25788 administered to the
animals was 0.36 mCi/mMole. Glass/stainless steel metabolism
cages housed the treated animals and allowed the collection
of exhaled CO_2 in 10% KOH solutions (3). Urine and feces
samples were frozen upon collection for later analysis.
Acetone solutions containing [^{14}C]R-25788 were
incorporated into Felton Loamy Sand soil (89% sand, 6% silt,
5% clay, 2% organic matter) as the soil was mixed in a small
cement mixer. Final treatment rate was 0.5 lb R-25788/acre
for plant metabolism studies and 1.0 lb/acre for soil
metabolism studies. In some tests, unlabeled EPTC
(Stauffer) herbicide was added at 6 lb/acre to the same soil
in a similar manner using dilutions of the 6E formulated product. DeKalb XL374 hybrid corn was seeded directly in 500
cm^3 paper cups filled with treated soil. To facilitate
metabolite isolation, corn seeds were germinated on moist
filter paper and then transferred to beakers containing
nutrient solution (4). After eight days in the nutrient
solution, the seedlings were treated with [2-^{14}C]R-25788
dissolved in nutrient solution. Two days after treatment,
seedlings were harvested for extraction.
Some plants were grown in metabolism chambers consisting of bell jars through which air was drawn into two
traps in series. The traps were an aqueous suspension of
charcoal followed by a 5% KOH solution. A second chamber
with a container of treated but unplanted soil served as a
control. Other plants in treated soil were grown in a green
house. Soil metabolism samples in glass pint jars were held
in a growth chamber at 27°C and the samples were maintained

at approximately 10% moisture level. Some soil samples
were treated aseptically with [^{14}C]R-25788 after the soil
had been autoclaved (1 hr., 120°C, 15 psi).

Urine samples and extracts of feces, plants and soils
were chromatographed on Analtech (Newark, DE) thin-layer
chromatography G or GF plates with either analytical (250
micron) or preparative (1000 micron) layers. Solvent
systems for tlc were benzene:ethyl ether 2:3 (BE), benzene:
hexane 1:1 (BH), benzene:chloroform 1:1 (BC), n-butanol:
ethanol:water 4:1:1 (BEW), n-butanol:acetic acid:water
4:1:1 (BAW) and chloroform:acetone methanol 5:4:1 (CAM).
Radioactive areas were located by autoradiography (Kodak
RP/R54 X-ray film) and were scraped from the supporting
glass plates for extraction or direct counting by lsc.

C. Metabolite Isolation

Fresh urine samples and acidified urine samples (pH 2)
were extracted with benzene and ethyl ether, respectively,
to remove non-polar metabolites and the remaining aqueous
samples were lyophilized to dryness and redissolved in
methanol. Plant tissue samples were extracted with ethanol:
water (7:3) and filtered to recover the pulp. The plant
extracts were partitioned between chloroform (organic
phase) and water after rotary evaporation to concentrate the
original extract. Soil samples were extracted for 8 hours
by Soxhlet extraction with acetone followed by 3 hours of
shaking in water. Filtered residues of the soil
were combusted for total ^{14}C. Extracted samples were
concentrated by rotary vacuum evaporation or lyophilization.

Preliminary purification of the aqueous phase
metabolites from the corn plant extracts was performed on a
2.5 by 98 cm Sephadex G-10 column (Pharmacia, Piscataway,
NJ) eluted with water. To purify some fractions,
chromatography columns packed with silicic acid were eluted
sequentially with benzene, ethyl ether, acetone and methanol.
Radiocarbon in the collected column fractions was determined
by lsc before fractions were pooled for further analysis.

Urine samples were subjected to enzymatic hydrolysis
using β-glucuronidase GL (Worthington, Freehold, NJ) and
β-glucuronidase/aryl sulfatase, Grade B (Calbiochem,
La Jolla, CA) in 0.05 M sodium acetate buffer adjusted with
HCl to pH 4.5 and pH 6.5, respectively. A typical incubation
had 100,000 dpm of urinary radiocarbon, 0.1 ml water, 0.3 ml
buffer, and 0.04 IU of β-glucuronidase or 800 Whitehead units
of aryl sulfatase. Control samples were run in pH 4.5
buffer with no added enzyme. Aliquots of test samples which
had been incubated for 18 hr at 37°C were chromatographed

directly. Preparative scale incubation with β-glucuronidase/
aryl sulfatase was used to liberate larger quantities of de-
conjugated metabolites. These were recovered from the
incubation mixture by ether extraction and were purified
by preparative tlc. Samples of [^{14}C]R-25788 plant
metabolites were hydrolyzed with 1 mg of β-glucosidase
(pH 5.0) at 37°C for 20 hr or with 1N HCl under reflux for
2 hr. The hydrolyzed material was extracted with diethyl
ether and assayed for ^{14}C prior to tlc for purification.

D. In Vitro Metabolites

Male rats (200-350g) were starved for 12-24 hours before
sacrifice. The livers were removed, pooled, and a 20 to 30%
homogenate made in cold pH 7.6 0.1M potassium phosphate
buffer. The microsomal fraction and soluble fraction were
obtained by ultracentrifugation of the postmitochondrial
supernatant at 100,000g for 1 hour. In some cases low
molecular weight species such as glutathione (GSH) were
removed from the soluble fraction by passage through a column
of Sephadex G-25.
Incubations were carried out in mini-culture tubes
(0.8 ml). The [^{14}C]R-25788 (5μl, final concentration 0.5 mM)
dissolved in chloroform was added to 5 μl distilled water and
the chloroform removed with a gentle stream of argon. Enzyme
(60μl) and cofactors (20μl) were added, and the reaction
tube was stoppered and shaken at 37°C. At various intervals
5μl aliquots were spotted on silica GF tlc plates. After
development, the radioactive areas were located by autoradiography, scraped from the plate, and quantified by
scintillation counting.

E. Gas Liquid Chromatography and Mass Spectrometry

The reagents boron trifluoride in n-butanol (Applied
Science, State College, PA) and Diazald® (Aldrich Chemical,
Milwaukee, WI) were used as described in the suppliers
literature. Gas liquid chromatography (glc) was done with
3% OV-17 or OV-1 on GC-Q (Applied Science) in 5 foot glass
columns in a Finnigan 9500 gas chromatograph (Finnigan Corp,
Sunnyvale, CA). The column outlet was split 1:10 for the
hydrogen flame detector and a radioactive monitoring system
(Selectra System 5000, Barber-Coleman, Rockford, IL),
respectively. A Finnigan 1015D gas liquid chromatography/
mass spectrometer (gc/ms) was used to identify metabolites
and the parent compound from their mass spectra.

III.   RESULTS AND DISCUSSION

A.   Balance Study and Tissue Residues in Rats

Rats (2 male and 2 female) dosed once with 75 mg/kg of [^{14}C]R-25788 rapidly excreted and exhaled the ^{14}C in urine, feces and exhaled air (Figure 2). Only small portions of the radiocarbon were excreted or exhaled after 48 hours. The presence of $^{14}CO_2$ in the exhaled air at early intervals indicates rapid and extensive metabolism of the [^{14}C]R-25788 since the labeled carbon atom has two chlorine atoms attached.

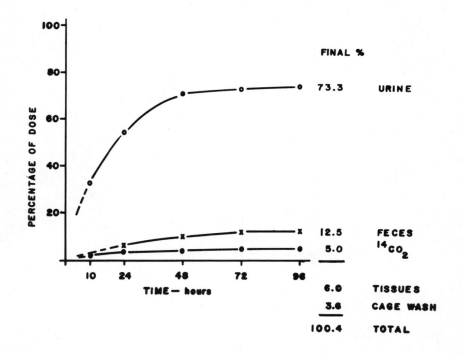

Fig. 2. Average excretion pattern and total tissue residues for four rats treated with 75 mg/kg of [^{14}C]R-25788 by oral intubation.

After 96 hours, the animals were sacrificed and their tissues were analyzed for ppm equivalents of R-25788 based on total ^{14}C content. An average of 6% of the radiocarbon dose was present in the tissues (Figure 2). Tissue residue analysis showed that organs involved in metabolism had the highest ppm values while the carcass, fat, gonads and brain

had the lowest residue values. Since [14]CO_2 was exhaled, it
is likely that the tissue residues at 4 days represent
some incorporation of [14]CO_2 into the intermediary metabolic
pools through a two carbon intermediate acid or aldehyde.
Total recovery of ^{14}C from urine, feces, CO_2 and tissues for
the 96 hour experimental period averaged 100.4% of the dose
(Figure 2).

B. Isolation of Urinary Metabolites

The benzene extract of 24-hour male rat urine contained
3.88% of the total urinary[14]C while 13.11% of the ^{14}C in
the female urine sample was extractable. Individual
metabolites in the benzene extracts were separated by tlc
in the BE solvent system to reveal one major metabolite and
several minor metabolites (Figure 3). The least polar

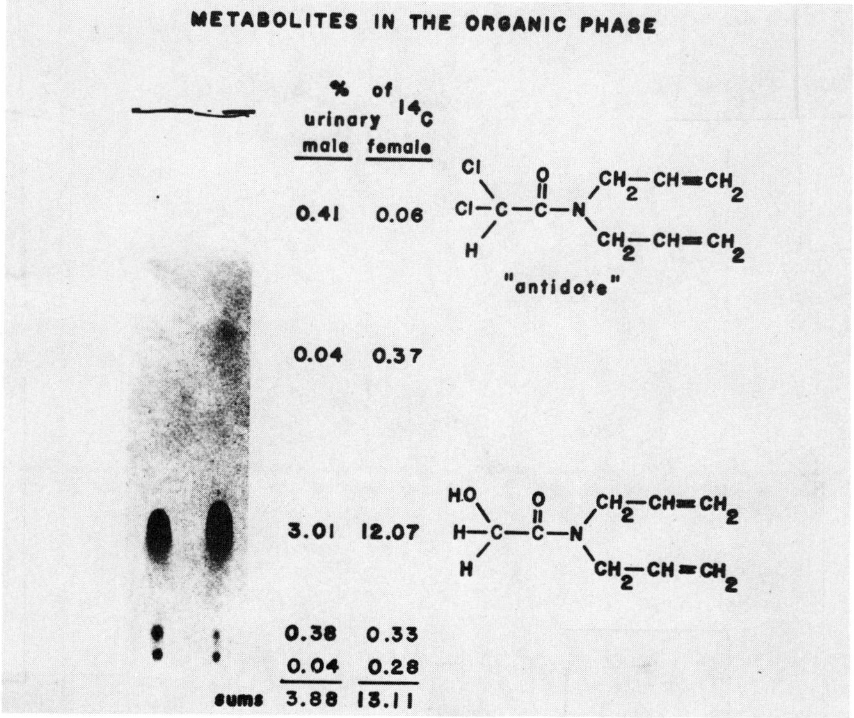

Fig. 3. Thin layer chromatography of the 24 hour
benzene extractable urinary metabolites of R-25788 separated
in the benzene:ether 2:3 solvent system and the distribution
of radioactivity.

component was isolated by preparative tlc in the BE solvent
system and identified as unchanged R-25788 by cochromatography with the original R-25788. The component and R-25788
had identical glc retention times (4.4 min) on the 3% OV-17
column at 150°C with 40 ml of helium per min. Male rats
excreted a greater proportion of the dose as unchanged
R-25788 than did females but the value was less than 0.5%
in both cases indicating rapid and extensive metabolism
of the antidote by the rats.

The major benzene extractable metabolite from both male
and female urine samples was isolated by preparative tlc and
found to elute in 3.4 min from the glc under the same
conditions used above while using the radioactive monitoring
system for detection of ^{14}C. The compound was identified
from its mass spectrum as N,N-diallylglycolamide (Figure 4).

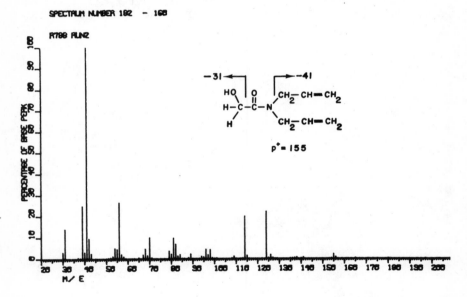

Fig. 4. Mass spectrum of the major benzene extractable
[^{14}C]R-25788 urinary metabolite, N,N-diallylglycolamide.

The major fragments are those remaining after the loss of
an allyl (114 m/e) or a methylol (124 m/e) radical. The
compound was synthesized and its glc and tlc characteristics
were found to be identical to that of the isolated metabolite.
When evaluated for antidotal activity in the established
screening procedure, the metabolite was inactive.

The majority of urinary metabolites were polar compounds
which did not extract into benzene but were separated by

chromatography in a BEW by BAW two-dimensional tlc system
(Figure 5). Of the five major ^{14}C labeled metabolites, only

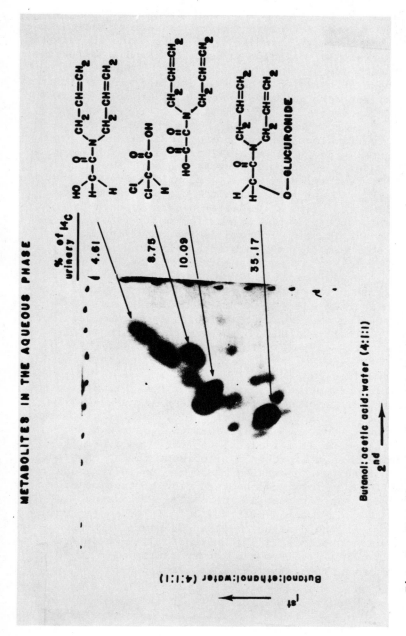

Fig. 5. Two-dimensional tlc of the water soluble 24 hour urinary metabolites of [^{14}C]R-25788 and their percent distribution.

the most polar compound, accounting for 35% of the urinary
^{14} C, was hydrolyzed by β-glucuronidase and the β-
glucuronidase/aryl sulfatase preparations. After preparative
deconjugation, the aglycone was identified as N̲,N̲-diallyl-
glycolamide by its tlc, glc and gc/ms characteristics. The
major metabolite was, therefore, the glucuronide conjugate
of the N̲,N̲-diallylglycolamide previously found as a benzene
soluble metabolite.

Upon acidification of the aqueous phase from the urine,
two metabolites were extracted into ethyl ether. The less
polar acid, representing 8.75% of the urinary radiocarbon,
was shown by two-dimensional tlc in BEW by BAW to be
identical to the standard dichloroacetic acid. This product
is formed upon hydrolysis of R-25788 in vivo. The second
acid (10.09% of the ^{14}C) was derivatized with BF$_3$ in n̲-
butanol. The esterified product was purified by preparative
tlc in the BH solvent system and the ester was identified
by gc/ms from its chemical ionization mass spectra in both
methane and isobutane. The identified metabolite was N̲,N̲-
diallyloxamic acid, now observed as its butyl ester. A
synthetic standard of this acid cochromatographed with the
isolated metabolite. When tested for antidotal activity,
the metabolite was inactive.

The least polar metabolite remaining in the urine after
benzene extraction was N̲,N̲-diallylglycolamide indicating
that benzene had not extracted all of this metabolite from
the urine. Other reference compounds which were not
metabolites of R-25788 in rats were the mono allyl analog
of R-25788 (N̲-allyl-2,2-dichloroacetamide), glycolic acid
and oxamic acid.

The proposed metabolic pathway for the antidote in rats
involves two distinct degradation routes (Figure 6). The
favored route is the formation of the N̲,N̲-diallylglycolamide
and its glucuronide conjugate. Formation of the N̲,N̲-
diallylglycolamide also leads to the observed N̲,N̲-diallyl-
oxamic acid which, if further metabolized, may account for
some ^{14}CO$_2$ found as an exhaled product. However, hydrolysis
of the R-25788 to form the observed dichloroacetic acid
probably accounts for most of the ^{14}CO$_2$. In a separate
experiment, a rat treated with [2-^{14}C]dichloroacetic acid
exhaled most of the dose as ^{14}CO$_2$ in 24 hours. The numerous
minor polar metabolites in the urine of ^{14}C R-25788 treated
rats presumably arise from the incorporation of ^{14}CO$_2$ or
its precursors in the intermediary metabolic pathways in
the rat.

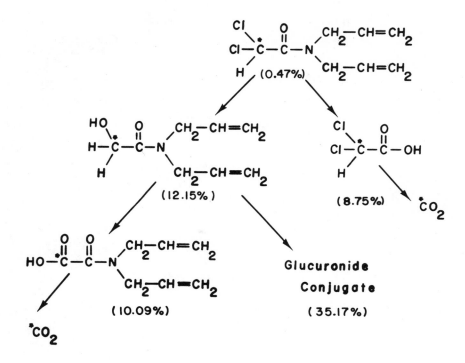

Fig. 6. Proposed metabolic pathway for $[^{14}C]R$-25788 in rats with the percent distribution of metabolites in the 24 hour urine samples.

C.   In Vitro Metabolism by Liver Fractions

A limited in vitro metabolism study has been conducted using $[^{14}C]R$-25788 incubated with rat liver microsomal and soluble fractions to examine the subcellular distribution of degradative enzymes and cofactor requirements for metabolism of R-25788. Although some metabolism occurred in the microsomal fraction when fortified with NADPH, the soluble fraction was considerably more active in the degradation of R-25788, and further studies were limited to this fraction, its products, and its cofactor requirements.

As shown in Table 1, R-25788 is stable under the reaction conditions in the absence of enzyme and cofactors. A slight but definite spontaneous reaction occurs between this substrate and GSH in the absence of enzyme leading to several unidentified polar compounds. NADPH and the other nucleotides tested do not react directly with the substrate.

TABLE 1

*Effect of Cofactors on the Metabolism of [^{14}C]R-25788
by Rat Liver Soluble Fractions[a]*

Incubation No.	1	2	3	4	5	6
Enzyme	-	-	+	+	+	+
GSH (5mM)	-	+	-	-	+	+
NADPH (2mM)	-	-	-	+	-	+
	Percentage of substrate recovered as:					
R-25788	99.0	97.1	97.8	94.1	63.5	56.1
'Aldehyde'	0.1	0.3	0.6	0.5	22.3	2.4
'Alcohol'	0.6	0.5	0.4	3.3	4.7	35.8
Polar metabolites[b]	0.2	1.9	1.0	1.6	8.2	4.4

a *Soluble fraction passed through Sephadex G-25,
concentration in incubation equivalent to 71 mg liver per
ml. Reaction time, 30 min. Metabolites separated on
silica gel GF by the CAM solvent system. R_f values were:
Polar metabolites, 0.00; 'Alcohol', 0.45; 'Aldehyde', 0.65;
R-25788, 0.78.*

b *Polar metabolites consist of several compounds including
N,N-diallyl oxamic acid, and GSH-derived conjugates.*

As long as GSH is removed from the soluble fraction, little
or no biotransformation occurs and 98% of the substrate can
be recovered unchanged. However, in the joint presence
of GSH and soluble fraction a rapid reaction ensues with the
loss of 37% of the substrate in 30 minutes and the
generation of several metabolites. Cochromatography with
authentic standards has allowed the tentative identification
of one major product as N,N-diallylglycolamide ('alcohol'
in Table 1). Among the highly polar products are N,N-
diallyl oxamic acid and probably one or more glutathione-
derived conjugates. Indirect evidence and chromatographic
behavior are consistent with a further major product being
N,N-diallylglyoxylamide ('aldehyde' in Table 1).
    In the absence of exogenous NAD nucleotides, the major
GSH-dependant product formed by the soluble fraction is
this aldehyde. The metabolic pattern is shifted in the
presence of NADPH (Table 1) or NADH to favor the related
alcohol at the expense of the aldehyde. In the presence
of either nucleotide with GSH, the total breakdown of

R-25788 is greater than with GSH alone. Neither nucleotide catalyzes significant metabolism of R-25788 by the soluble fraction in the absence of GSH.

The reduction of the aldehyde to the alcohol is stimulated equally by NADPH or NADH. Conversion in the former case is inhibited 63% by 2.5mM phenobarbital, an inhibitor of the NADPH-dependant aldehyde reductase (5). With NADH as cofactor, 80% inhibition of reduction is observed with a mixture of isobutyramide/pyrazole (12.5mM) which inhibits the NADH-dependant alcohol dehydrogenase (6). Thus it seems that both enzymes can carry out this aldehyde to alcohol reduction.

In these studies, and companion ones with unlabeled antidote and ^{35}S-GSH, only small amounts of glutathione-containing conjugates could be isolated even when rapid GSH-dependant metabolism was occurring. Thus, if the major route of metabolism is through a glutathione conjugate, it is an unstable one under the reaction conditions.

Based on these preliminary results, the scheme in Figure 7 is proposed for the reactions carried out by the liver soluble fraction. An unstable glutathione conjugate is formed initially by glutathione transferase. It is likely but unproved that this occurs by removal of a labile chlorine atom. A comparable reaction involving dechlorination by glutathione transferases in rats and plants has been reported in the metabolism of the related monochloroacetamide herbicides propachlor (2-chloro-N-isopropyl-acetanilide) and CDAA (N,N-diallyl-2-chloroacetamide) (7). The glutathione conjugate of R-25788 breaks down by a sequence not currently known to yield the aldehyde which is interconvertible to the related alcohol and acid by NAD(P) linked dehydrogenases and reductases. The latter two compounds and their conjugates are the major urinary metabolites found after oral administration of R-25788 to rats.

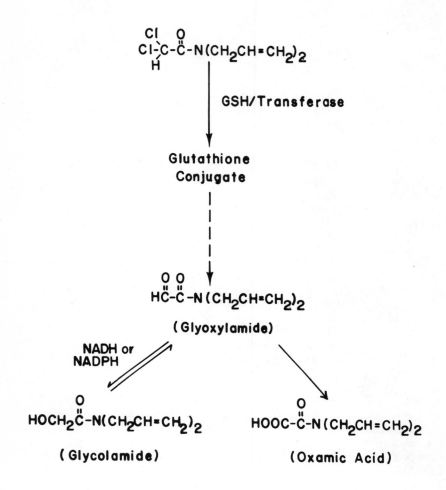

Fig. 7. *Proposed pathway of metabolism of* [^{14}C]*R-25788 by soluble fraction of rat liver.*

D. Corn Plant Metabolism of [^{14}C]R-25788

The uptake of radioactivity from corn shoots grown in [^{14}C]R-25788 treated soil was measured at several intervals between 4 and 10 days by combustion and lsc analysis for total ^{14}C in the plant shoots (Figure 8). Uptake was nearly linear for the 10 days and no differences in the results were observed when EPTC herbicide was also

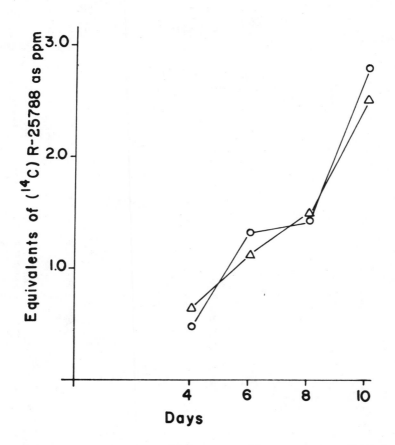

Fig. 8.   Uptake of ^{14}C *from* $[^{14}C]R$-*25788 treated soil (0.5 lb/acre) by corn shoots expressed as ppm equivalents of R-25788.*

Treatment: △ $[^{14}C]R$-*25788 at 0.5 lb/acre;* ○ $[^{14}C]R$-*25788 at 0.5 lb/acre plus EPTC herbicide at 6 lb/acre.*

incorporated at the recommended rate of 6 lb/acre. In another experiment, extraction of 4, 7, 10 and 14 day old plant shoots grown in R-25788 treated soil or 7-day old shoots grown in antidote plus EPTC treated soil all resulted in nearly the same 73% of the radiocarbon as water soluble products, 16% as $CHCl_3$ soluble products and 11% unextractable ^{14}C. The metabolites in the chloroform and aqueous phases were separated best by tlc in the BE and BAW tlc solvent systems, respectively (Figure 9).

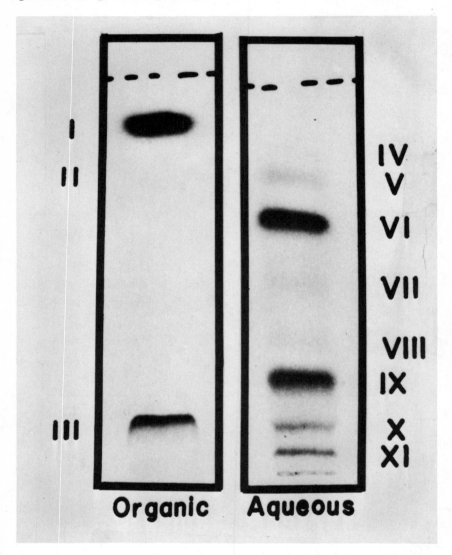

Fig. 9. Tlc of corn shoot organo and aqueous phase metabolites separated in the BE (organo) and BAW (aqueous) solvent systems.

Use of nutrient solution treatment of corn plants resulted in 4 to 10 fold increases in the total ^{14}C uptake as compared to soil grown plants but chromatography of the metabolites from plants grown in nutrient solution and soil showed a different distribution of identical metabolites. For convenience of isolation further work to identify the metabolites was done with plants grown in nutrient solution but the distribution of metabolites reported is based on soil grown plants.

The organo soluble metabolites were purified by repeated preparative tlc in the BE and BC solvent systems. Using the OV-17 glc column for final separation, the metabolites designated I, II and IV were identified from their mass spectra as R-25788, N-allyl-2,2-dichloroacetamide and N,N-diallylglycolamide, respectively.

The water soluble metabolites were separated by the BAW tlc solvent system into two major and several minor components (Figure 9). Metabolite VI extracted into ether upon acidification (pH 2) of the aqueous phase and was eluted from a silicic acid column in the methanol fraction. The compound was further purified by preparative tlc in BAW and finally methylated for gc/ms analysis. Its mass spectrum although contaminated with fragments of unrelated components, showed the metabolite VI to be N,N-diallyloxamic acid, observed as its methyl ester (Figure 10).

Metabolites IV and V were also extracted into ether upon acidification of the aqueous phase. Metabolite V was identified by cochromatography in two dimensional BEW by BAW as dichloroacetic acid after purification by preparative tlc. Upon enzymatic (β-glucosidase) hydrolysis, component IV liberated N,N-diallylglycolamide and was, therefore, designated as the glucoside of N,N-diallyl-glycolamide. Metabolite IX was subjected to preparative tlc (BEW) purification and its tlc migration suggested that it was oxalic acid, but cochromatography was unsuccessful and its structure is unknown.

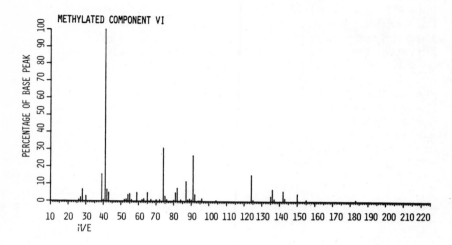

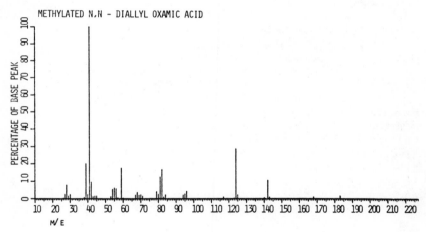

Fig. 10. Mass spectrum of methylated corn plant metabolite VI and methylated N,N-diallyl oxamic acid.

To study the possible liberation of $^{14}CO_2$ from the corn plants, some plants were grown in the bell jars with a treated but unplanted soil sample serving as control (Figure 11). The charcoal and KOH traps for both the planted and unplanted soil contained significant radioactivity. After correcting for the ^{14}C from the unplanted soil, it was found that the corn seedlings grown in[^{14}C]R-25788 treated soil liberated approximately 6% of the total radioactivity taken up by the plants as $^{14}CO_2$ and volatile products. Due to the facile nature of the reaction in soil

Fig. 11. Plant metabolism chambers for recovery of volatile products and CO2 from plants and balanced with an unplanted but treated soil sample.

to liberate $^{14}CO_2$ from $[^{14}C]$ R-25788 (see next section) further tests of the ability of corn plants to release $^{14}CO_2$ were not conducted.

The proposed metabolic pathway for R-25788 in corn plants is qualitatively identical to that found in rats except that the corn plants produced a small quantity of N-allyl-2,2-dichloroacetamide which was not present in rat urine sample (Figure 12). The principal route of

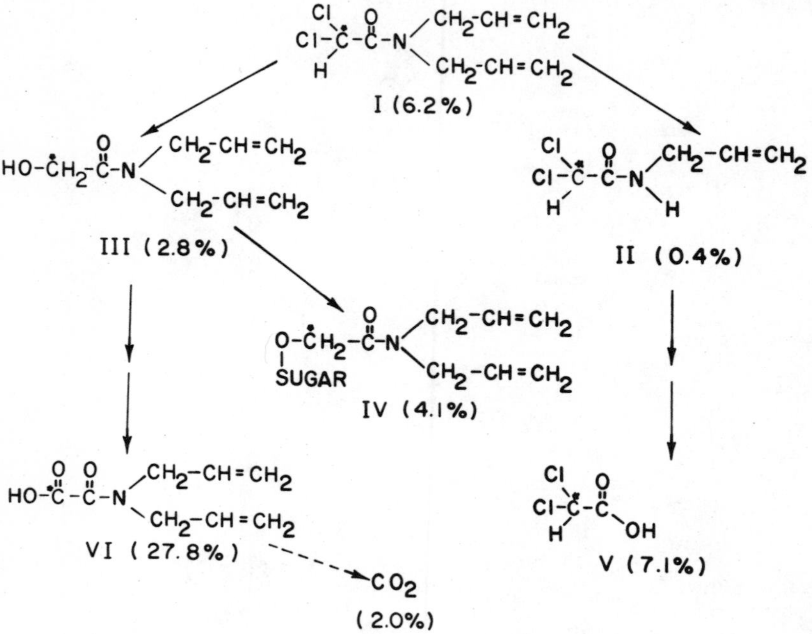

Fig. 12. Proposed metabolic pathway for [^{14}C]R-25788 in corn plants. Percent figures are for total ^{14}C in corn shoots.

degradation in corn results from replacement of the two chlorine atoms in R-25788 probably forming N,N-diallyl-glycolamide which accounts for 3% of the extractable radio-activity in the corn shoots. The glycolamide is the likely intermediate in the formation of the major identified metabolite, N,N-diallyloxamic acid (28% of the ^{14}C). An alternate fate for the N,N-diallylglycolamide is its conjugation to form a glycoside (4% of the ^{14}C). Another degradation route is the N-dealkylation of the R-25788 which accounted for 0.4% of the ^{14}C. Hydrolysis of R-25788 or the N-allyl-2,2-dichloroacetamide leads to the observed dichloroacetic acid (7% of the ^{14}C). Finally, 6% of the extractable radiocarbon in the corn shoots was unchanged R-25788. Of the total ^{14}C in the plant 48.5% is accounted for by the identified metabolites while unextractable residues in the pulp accounted for 11% of the ^{14}C, and the

other 6 major and minor metabolites make up the difference
of 40.5%.

E.   Soil Metabolism and Half Life Studies

The acetone extraction of the soil samples resulted in
quantitative removal of $[^{14}C]$R-25788 from both soil types
at zero time (Figure 13).   The samples extracted at later

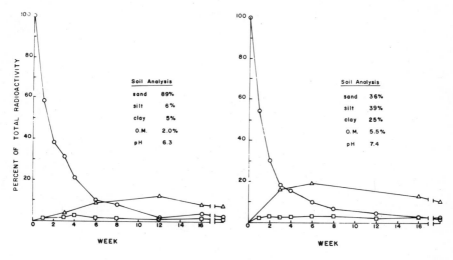

Fig. 13.   Percent distribution of acetone soluble (O),
water soluble (□) and unextractable (△) radioactivity from
$[^{14}C]$ R-25788 treated Felton loamy sand (left) and Sorrento
loam (right) soils treated at 1.0 lb/acre.

intervals showed that the radioactive acetone extractable
material was rapidly decreasing in both soil types while
water soluble metabolites accounted for small portions of
the radioactivity at each time interval.   By combustion of
the extracted soil residues, it was shown that the unextractable radiocarbon had increased in the loam soil to a maximum
of 19% of the original material at 6 weeks while the loamy
sand soil showed a 11% maximum at 12 weeks.   Both soils
showed decreases in unextractable materials at later
intervals.   The total ^{14}C found in all the fractions
decreased rapidly from the original level.   When antidote
treated loam soil was held in the bell jar metabolism
chamber for 21 days, 38% of the ^{14}C was lost through
volatility either as $^{14}CO_2$ (26%) or untrapped volatile
products (12%) indicating the soil rapidly decomposed the

antidote. The role of microorganisms in the soil metabolism
of the antidote was studied using the sterile and non-sterile
soil samples. The results show that most of the degradation
of R-25788 is the result of the microbial action.

The radioactive components in the organic phases from
the soil extracts were separated by tlc in the BE solvent
system (Figure 14). Most of the ^{14}C was present as R-25788

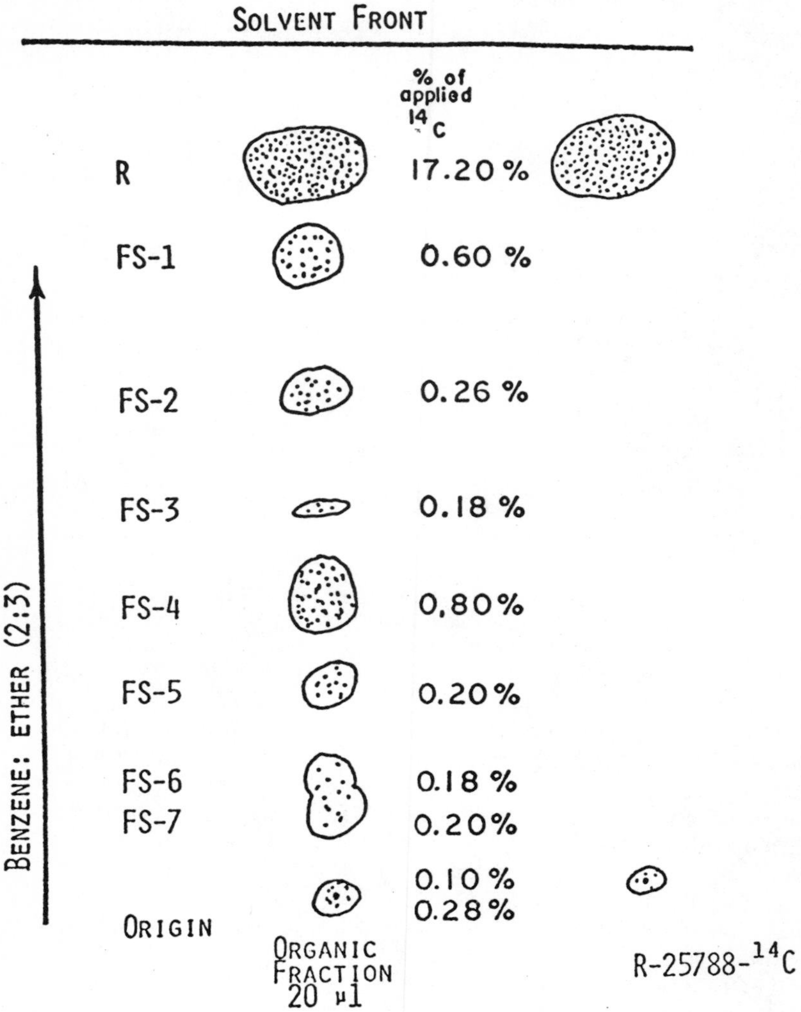

Fig. 14. *Facsimile of a tlc autoradiograph from the
concentrated acetone extract of soil 1 month after treatment
with* [^{14}C]*R-25788.*

at every interval but 7 minor metabolites were also present
in the extracts. These metabolites were isolated and
identified by cochromatography and gc/ms analysis. Metabolite
1 was resolved into two components which were identified as N-
allyl-2,2-dichloroacetamide and N,N-diallyl-2-chloroacetamide
while metabolite 4 was identical to N,N-diallylacetamide.
The finding of the reductive dechlorination to the N,N-
diallyl-2-chloroacetamide is unique to the soil metabolism
study since this product was not observed as a rat or corn
plant metabolite of [^{14}C] R-25788.

In conclusion, the studies reported here have shown that
R-25788 was metabolized by rats, corn plants and soils
through similar pathways. The primary pathway involves
removal of both chlorines from the R-25788 to form the
substituted glycolamide and oxamic acid metabolites. It has
been reported that an enzyme from pseudomonad isolated on
dichloroacetate as a carbon source carries out the analogous
reaction in which a hydroxyl group from water replaces the
first chlorine in dichloroacetate followed by an enzymatic
or spontaneous loss of the second to form the observed
glyoxylate (8).

IV. REFERENCES

1.   Gray, R. A., and Joo, G. K., in "Chemistry and
     Action of Herbicide Antidotes" (F.M. Pallos and J. E.
     Casida, Eds.), p. 67 . Academic Press, New York, 1978.
2.   Jaworski, E. G., J. Agric. Food Chem. 12, 33 (1964).
3.   Ford, I. M., Menn, J. J., Meyding, G. D., J. Agric.
     Food Chem. 14, 83 (1966).
4.   Hoagland, D. R., and Arnon, D. I., The Water-Culture
     Method for Growing Plants Without Soil, Calif. Agr.
     Expt. Sta. Cir. 347, 39 (1938).
5.   Ris, M. M., and von Wartburg, J., Europ. J. Biochem.
     37, 69 (1973).
6.   Branden, C., Jörnvall, H., Eklund, H., and Furugren, B.,
     Alcohol Dehydrogenases, in "The Enzymes" Vol. XI,
     Part A., 3rd Ed. (P. D. Boyer, Ed.) pp. 103-190.
     Academic Press, New York, 1975.
7.   Lamoureux, G. E., and Davison, K. L., Pesticide Biochem.
     Physiol. 5, 497 (1975).
8.   Goldman, P., Milne, G. W. A., and Keister, D. B.,
     J. Biochem. 243, 428 (1968).

GLUTATHIONE CONJUGATION: A MECHANISM FOR HERBICIDE DETOXICATION AND SELECTIVITY IN PLANTS

Richard H. Shimabukuro, Gerald L. Lamoureux, and
D. Stuart Frear
Metabolism and Radiation Research Laboratory
Agricultural Research Service
U. S. Department of Agriculture

Glutathione conjugation of herbicides appears to be a common reaction in plants and mammals. However, the terminal product in plants is not the mercapturic acid formed in mammals. In plants, glutathione conjugation is a detoxication mechanism which effectively reduces the active internal concentration of the herbicide and prevents irreversible injury.

I.   INTRODUCTION

Several extensive reviews on herbicide metabolism, mode of action, biochemistry and physiology have recently appeared (1, 2). These reviews indicate that much is known on the behavior and fate, mode of action, and selectivity of herbicides in plants. It is apparent from these reviews that research on the selective action of herbicides has focused primarily on environmental and ecological factors, formulation technology, inherent morphological differences and absorption and translocation as principal factors in selectivity. All of these factors undoubtedly contribute to the selective action of herbicides. However, the rapid advances made within the past few years on herbicide metabolism in plants indicate clearly that the most important single factor contributing to selectivity may be the ability of a plant to metabolize and detoxify the herbicide. Herbicide biotransformation reactions in plants include oxidation, reduction, hydrolysis and conjugation with natural products as principal metabolic reactions. The conjugation of herbicides with glutathione (GSH) is rapidly becoming established as one of the major detoxication and selectivity factors in plants. Recently, the mode of action of the herbicide antidote, N,N-diallyl-2,2-dichloroacetamide, was reported to be linked to glutathione conjugation (3). This discussion reviews glutathione conjugation of herbicides in plants and the significance of this reaction to herbicide detoxication and selectivity.

II. MERCAPTURIC ACID BIOSYNTHESIS IN MAMMALS

Glutathione conjugation and the formation of mercapturic acids from xenobiotic compounds have been investigated extensively in mammals. Several reviews on GSH conjugation and mercapturic acid formation have appeared (4, 5). The formation of mercapturic acids in mammals is a detoxication mechanism where non-polar foreign compounds are converted to more hydrophilic forms for elimination from the body. Fig. 1 shows the four steps in the formation of mercapturic acids. The initial step involves an enzyme-catalyzed nucleophilic displacement reaction to give a GSH conjugate. Sequential removal of the glutamic acid and glycine residues, respectively, yields the S-cysteine conjugate which is N-acetylated in the final step to give the mercapturic acid. This product is excreted in the urine of mammals. The initial step   (conjugation with GSH) appears to be the crucial step in the destruction of biological activity of the foreign compound.

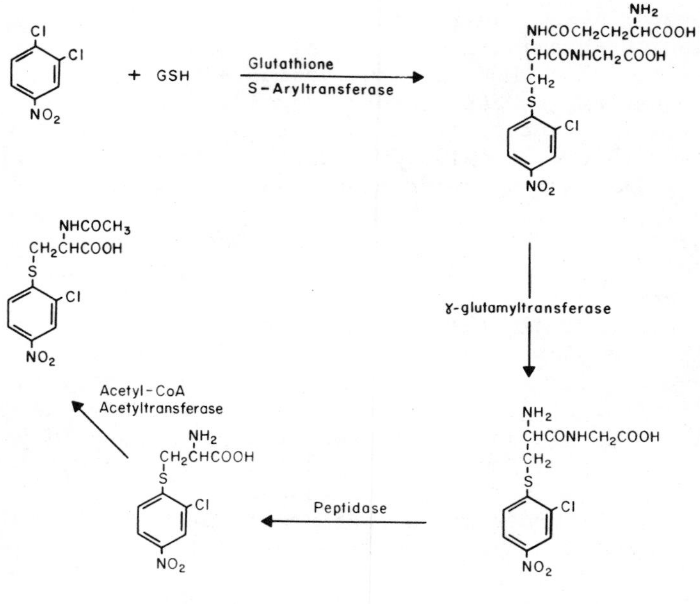

Fig. 1.   Biosynthesis of mercapturic acids.

Conjugation with GSH is catalyzed by glutathione S-transferases with different substrate specificities (4, 5). In plants only two glutathione S-transferases have been isolated and characterized (6, 7). These enzymes are discussed in this review in relation to herbicide selectivity.

III. GLUTATHIONE CONJUGATION IN PLANTS

The number of herbicides known to be metabolized by GSH conjugation in plants and mammals is very limited (5, 8). Generally, the initial reaction of the herbicides with GSH occurs in both plants and animals. However, limited studies indicate that the terminal products in plants are not the same as the mercapturic acids excreted by mammals. This difference between plants and animals may be due to the fact that plants do not have a system equivalent to the renal excretion system in mammals.

GSH conjugation of atrazine (2-chloro-4-ethylamino-6-isopropylamino-s-triazine) was the first report of GSH conjugation of a herbicide in plants (Fig. 2) (9, 10, 11). In mammals, GSH conjugation of atrazine and its N-dealkylated metabolites was demonstrated only in vitro (12). In sorghum, the GSH conjugate of atrazine was catabolized to the S-cysteine conjugate by an initial loss of the glycine followed by the loss of glutamic acid. The S-cysteine conjugate was not acetylated to yield the mercapturic acid as would be expected in mammals, but a non-enzymatic rearrangement to the N-cysteine conjugate occurred. Some dimerization resulted but the predominant reaction was the formation of the lanthionine conjugate by an unknown mechanism (10). Subsequent reactions leading to insoluble residues are unknown. It should be noted that the order of removal of the glycine and glutamate residues observed in sorghum was the inverse of that observed in mercapturic acid formation in mammals shown in Fig. 1.

Neither atrazine (13) nor simazine [2-chloro-4,6-bis-(ethylamino)-s-triazine] (5) yielded mercapturic acids when fed to rats. Cyprazine (2-chloro-4-cyclopropylamino-6-iso-propylamino-s-triazine) was metabolized in rats by hydrolysis of the 2-chloro substituent, N-dealkylation, and oxidation of the side chains (14). However, the GSH conjugate of cyprazine was found in several plant species (15). Hydrolysis of 2-chloro-s-triazines is also a major detoxication mechanism in corn while N-dealkylation is a minor detoxication mechanism present in most plants.

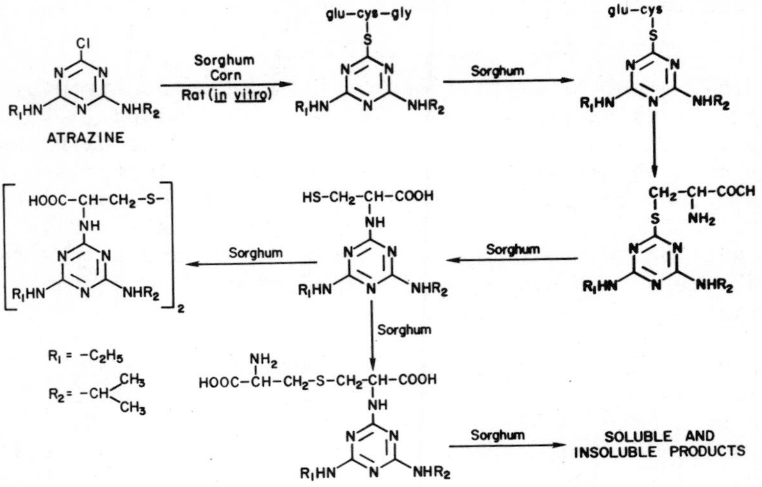

Fig. 2. *Metabolism of atrazine in plants and mammals.*

In contrast to atrazine, simazine, and cyprazine, cyanazine [2-chloro-4-ethylamino-6-(1-cyano-1-methylethyl-amino)-s-triazine] was metabolized by GSH conjugation to its mercapturic acid in the rat (Fig. 3) (16, 17). The GSH conjugate of cyanazine was not detected in soil and in corn plants grown in cyanazine-treated soil (18). However, several plant species formed the GSH conjugate of cyanazine when the herbicide was absorbed from nutrient solution (19). The broken arrows in Fig. 3 represent multiple reactions.

The metabolism of atrazine and other s-triazines indicates that a given herbicide or other closely related herbicides may or may not be metabolized by the mercapturic acid or mercapturic acid-like pathway in mammals and plants, respectively. The chemical properties of each herbicide appear to determine whether GSH conjugation will be an important reaction in either mammals or plants or in both as other examples indicate.

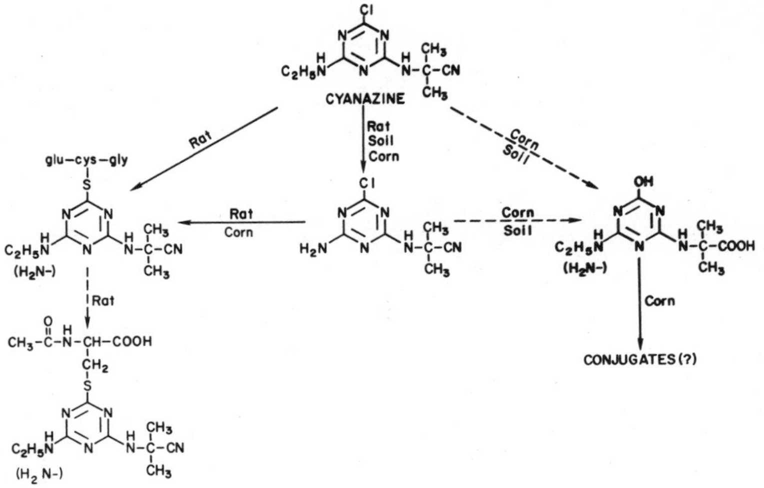

Fig. 3.   *Metabolism of cyanazine in plants, mammals, and soil.   Broken lines represent multiple reactions.*

GSH conjugation and catabolism of fluorodifen (2,4'-dinitro-4-trifluoromethyl diphenylether) occur in both mammals (20) and plants (21, 22) (Fig. 4).   The GSH conjugate resulting from GSH-dependent cleavage of the diphenylether was isolated from peanut (21) and its mercapturic acid was isolated from the urine of rats dosed with fluorodifen (20). The compounds in brackets (Fig. 4) are intermediates in the mercapturic acid pathway which have not been isolated from both peanut and rat.   Unlike atrazine in sorghum (Fig. 2), the S-cysteine conjugate of fluorodifen in peanut did not undergo rearrangement and metabolism to the lanthionine conjugate.   The S-cysteine conjugate was instead acylated to yield S-(2-nitro-4-trifluoromethylphenyl)-N-malonyl-cysteine (22).   This metabolite may be an intermediate in the formation of insoluble residue.

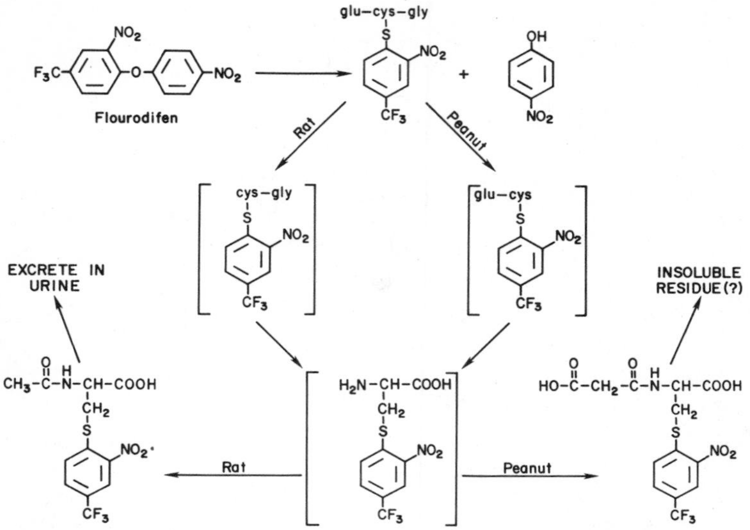

Fig. 4.   Metabolism of fluorodifen in plants and mammals.

Other examples of herbicides metabolized by GSH con-jugation are shown in Fig. 5.   The GSH and dipeptide conjug-ates of propachlor (2-chloro-N-isopropyl-acetanilide) have been isolated and characterized in plants (23).   The GSH conjugates of CDAA (N,N-diallyl-2-chloroacetamide) (23) and barban (4-chloro-2-butynyl-m-chlorocarbanilate) (23, 24) have been detected in several plant species.   These intermediates are presented in brackets (Fig. 5) because they have not been characterized by means other than thin-layer chromatography. The mercapturic acids of propachlor and CDAA have been identified in the rat (20).   The metabolism of fluorodifen, propachlor and CDAA indicates that GSH conjugation is a significant reaction in their catabolism in both mammals and plants.   However, the end products of GSH conjugation appear to be different in mammals and plants.

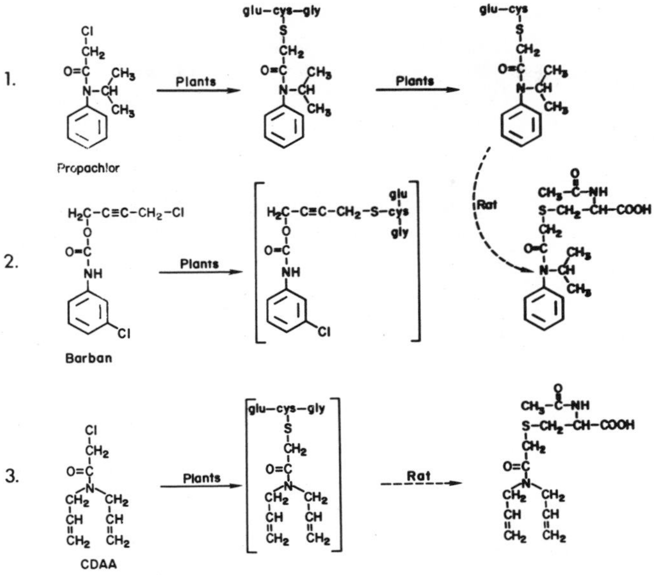

Fig. 5. Metabolism of propachlor, barban, and CDAA in plants and mammals. Broken lines represent multiple reactions.

GSH conjugation has been implicated in the mode of action of the herbicide antidote, R-25788 (N,N-diallyl-2,2-dichloroacetamide) against EPTC (S-ethyl dipropylthiocarbamate) injury in corn (3,25). The increased GSH conjugation of EPTC sulfoxide (Fig. 6) produced by R-25788 was reported to be a detoxication mechanism to protect corn against injury by EPTC (25). S-Carbamyl-mercapturic acid (S-(N,N-dipropyl-carbamyl)-N-acetylcysteine) was isolated from the urine of rats dosed with EPTC or EPTC sulfoxide (3, 26) but treated corn did not form this compound (25).

The limited number of examples discussed above indicates that GSH conjugation of herbicides and catabolism to mercapturic acids appear to be constant in mammals. In plants, the initial GSH conjugation of the different herbicides appears to be a common reaction, but the end products of catabolism are not mercapturic acids. Even the end products in plants may differ depending on the compound involved as indicated in the metabolism of atrazine and fluorodifen.

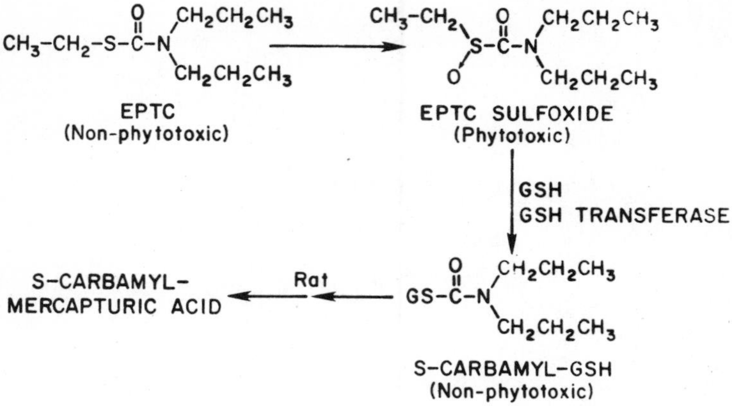

Fig. 6. GSH conjugation of EPTC sulfoxide as
influenced by R-25788.

IV.   GLUTATHIONE-S-TRANSFERASES

Several different types of mammalian glutathione S-
transferases have been described (4, 27). In plants only two
glutathione S-transferases have been characterized (6, 7).
The properties of these two enzymes are summarized in Table 1.
Both of these enzymes and all mammalian glutathione S-transferases are soluble enzymes. The differences in substrate
specificity, pH optima, distribution within the plant
source, and differences in the plant species containing these
enzymes indicate that these enzymes are not the same.
Atrazine was not a substrate for the enzyme isolated from
pea (7).   Pea and soybean are relatively susceptible to
atrazine injury and neither species metabolizes atrazine at
any appreciable rate to its water-soluble GSH conjugate.
However, both species contain an active glutathione S-transferase for fluorodifen metabolism.   Corn contains both of
these enzymes. However, the enzyme which catalyzes the GSH
conjugation of atrazine is active only in the shoots but the
corresponding enzyme for fluorodifen conjugation is distrib-
uted throughout the plant (6, 7).

TABLE 1

Characterization of Plant Glutathione S-Transferases

Properties	Substrate	
	Atrazine	Fluorodifen
Plant source	Corn	Pea
pH optimum	6.6 – 6.8	9.3 – 9.5
Km (substrate)	3.7×10^{-5}M	1.2×10^{-5}M
Km (GSH)	2.4×10^{-3}M	7.4×10^{-4}M
Distribution in plant	Shoots only	Shoots, roots, hypocotyls, etc.
Other plant sources	Sorghum, sugar cane, johnson grass	Peanut, corn, okra, cotton, soybean
Stability	Very stable	Very stable

The glutathione S-transferases from corn and pea were specific for reduced glutathione. Other sulfhydryl-containing compounds such as L-cysteine, mercaptoethanol, dithiothreitol and 2,3-dimercaptopropanol were ineffective substrates or inhibitors except for 2,3-dimercaptoethanol which strongly inhibited the enzymes from both plant sources (6, 7). The corn enzyme required substituted s-triazines with a chlorine in the 2-position and N-alkyl side chains in the 4- and 6-positions of the heterocyclic ring for enzyme activity. However, in vivo metabolism of 2-chloro-4-amino-6-isopropylamino-s-triazine to the GSH conjugate in sorghum indicated that both N-alkyl side chains may not be required for enzyme activity (10). Structural analogs of fluorodifen such as nitrofen (2,4-dichlorophenyl-p-nitrophenylether) and aminofluorodifen (2-amino-4'-nitro-4-trifluoromethyldiphenyl ether) were not substrates for the pea enzyme. However, these analogs were effective competitive inhibitors of fluorodifen GSH conjugation (7).

V.   GSH CONJUGATION AND HERBICIDE SELECTIVITY

Herbicide selectivity is a relative matter and is directly related to the dosage applied to plants. Any herbicide may be either phytotoxic or nonphytotoxic to all weeds or crop plants if applied at sufficiently high or low concentrations, respectively. However, at a reasonable physiological

concentration, some plants will be killed while others
remain uninjured. It is evident that herbicides may act on
more than one biochemically or physiologically sensitive site
(2). To be effective the herbicide must reach the sensitive
site or sites of action in its toxic form, at a concentration
sufficient to cause severe primary and secondary effects
which result in injury or disruption of normal growth
(Fig. 7). Injury may be only transient and plants may
recover or injury may exceed a threshold level where irreversible damage results and death becomes inevitable. Many
barriers such as absorption by roots or leaves, translocation,
spray retention, inherent morphological variations, etc. (2)
are believed to influence the active internal concentration
of a herbicide. However, investigations on mechanism of
herbicide selectivity indicate, that the ability of a species
to metabolize and detoxify a potential herbicide (1) is
probably the most important and effective means by which
plants reduce the concentration of the toxic herbicide to
prevent irreversible damage.

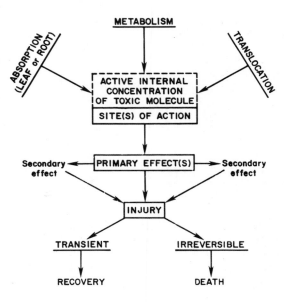

*Fig. 7. Factors affecting the mode of action and
selectivity of herbicides.*

GSH conjugation is a detoxication mechanism that effect-ively protects plants such as corn and sorghum (28) and other species (29) from atrazine injury. The correlation between GSH conjugation and recovery from atrazine injury with time may be readily demonstrated. Such a correlation may be difficult to demonstrate with herbicides such as propachlor, CDAA, EPTC or barban. Unlike atrazine, these herbicides have a different mode of action which makes it difficult to con-tinuously monitor the level of injury in response to GSH conjugation. However, GSH conjugation appears to be a major selectivity factor with the four herbicides mentioned above.

Atrazine is known to cause death in plants by inhibiting photosynthesis (1). Irreversible damage caused by primary and secondary effects of photosynthetic inhibition may be readily prevented in susceptible soybean by supplying an exogenous source of sucrose (30). Under natural conditions, plants may be expected to survive if photosynthetic in-hibition is not prolonged and recovery of normal photo-synthesis is achieved within a reasonable period of time. Injury under these circumstances would only be transient and not irreversible. This principal is illustrated in the results from atrazine-treated leaf discs of resistant sorghum and susceptible pea (Fig. 8) (31). Leaf discs were incubated in [^{14}C]atrazine for 1-1/2 hours, rinsed free of atrazine and oxygen evolution was measured over a 6-hour period. Atrazine metabolism was also determined after incubation in [^{14}C]atrazine and at the end of the 6-hour period. In resistant sorghum, a nearly normal rate of photosynthesis was achieved within 6 hours after a strong initial inhibition. This was accomplished by reducing the active internal concentration of atrazine from approximately 76% of all radioactivity in leaf discs at 1-1/2 hours to about 20% after 6 hours. Atrazine was converted primarily to its GSH conjugate over this period of time. Susceptible pea showed total inhibition of photosynthesis with no detox-ication of atrazine. In whole plants, prolonged photo-synthetic inhibition in pea could lead to unknown secondary effects that may cause irreversible injury and death as in atrazine-susceptible soybean (30).

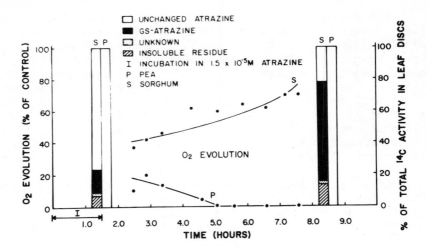

*Fig. 8. The response of photosynthesis to GSH
conjugation of atrazine in leaf discs of resistant
sorghum and susceptible pea.*

Leaf discs from atrazine resistant (GT112 RfRf) and
susceptible (GT112) isogenic lines of corn gave similar
results as described later (Table 2) (32). Photosynthesis
in the resistant corn line was nearly normal at all time
periods. Photosynthesis in the susceptible corn line was
still inhibited by 62% after 6.5 hours. The metabolism of
atrazine in the corn leaf discs indicated that the active
internal concentration of atrazine was reduced rapidly by
conversion to its GSH conjugate in the resistant line but not
in the susceptible line (Table 3). Therefore, metabolism of
atrazine by GSH conjugation occurred concomitantly with the
recovery from transient atrazine injury. The results with
leaf discs were similar to whole plant metabolism studies
(32).

TABLE 2

Atrazine-Inhibited Photosynthesis in Leaf Discs of Resistant and Susceptible Corn Lines

| Time* (hr) | Oxygen Evolution (% of control) | |
	GT112 RfRf (Resistant)	GT112 (Susceptible)
3.0	93.0	73.2
3.5	91.4	35.1
4.5	92.1	35.5
5.0	90.1	33.8
6.0	92.5	36.4
6.5	93.3	37.9

*Includes 1.5 hr incubation in 40 µM atrazine.

TABLE 3

Atrazine Metabolism in Leaf Discs of Resistant and Susceptible Corn Lines

| Inbred Line | Time* (hr) | Distribution of Extracted ^{14}C-Activity (%) | |
		Atrazine	GS-Atrazine
GT112 RfRf (Resistant)	1.5	82.7	17.3
	2.0	56.9	43.1
	2.5	45.4	54.6
	4.0	31.6	68.4
GT112 (Susceptible)	1.5	99.7	0.3
	2.0	99.6	0.4
	2.5	99.8	0.2
	4.0	99.4	0.6

*Includes 1.5 hr incubation in 37 µM [^{14}C]atrazine.

GSH conjugation converts atrazine from a small, hydrophobic, neutral molecule to a large, hydrophilic molecule with amphoteric properties. Such drastic changes in the physical properties alone may destroy the phytotoxic properties of the parent molecule (5). These changes undoubtedly will affect the ability of the parent compound to penetrate biological membranes such as the chloroplast membrane that surrounds the atrazine-sensitive site (31). Comparable changes in physical properties also occur when other triazines, propachlor, CDAA, barban, EPTC and fluorodifen are conjugated with GSH.

GSH conjugation of atrazine is an important mechanism for detoxication in corn. However, unlike sorghum, corn has another major detoxication mechanism that plays a significant role in its resistance to atrazine. Figure 9 presents a diagrammatic summary of atrazine metabolism and selectivity in corn (32). Atrazine is detoxified non-enzymatically to hydroxyatrazine and enzymatically to the GSH conjugate. N-Dealkylation is also a minor detoxication mechanism. The enzyme, glutathione S-transferase, is active in the leaves but not in the roots. The size of the arrows reflects the relative importance of the three detoxication mechanisms in corn. Hydroxylation appears to be the primary mechanism for resistance when atrazine is applied to the soil and the primary mode of entry into the plant is through its root. GSH conjugation is the major detoxication mechanism when atrazine is applied to the foliage. However, with only one exception several genetic lines of corn were all resistant to atrazine even with low hydroxylation activity in their roots. Unhydrolyzed atrazine was rapidly detoxified by GSH conjugation when translocated from roots to shoots. The example with corn illustrates the principal that herbicide metabolism may differ between plant tissues and that the sensitive sites of action may not necessarily be in close proximity to the site of detoxication. Also, the mode of entry of the herbicide into the plant may influence the products formed in its metabolism.

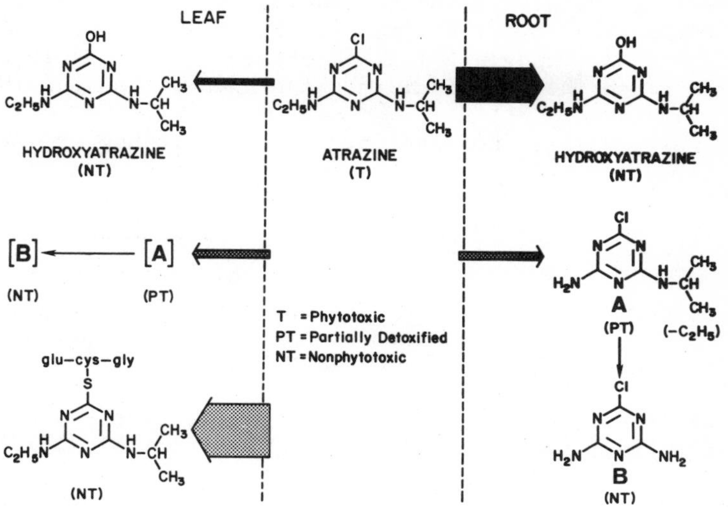

Fig. 9. Atrazine detoxication mechanisms and selectivity in corn.

In summary, plants metabolize several major herbicides by a mercapturic acid-like pathway. The initial GSH conjugation reaction appears to be common to both plants and mammals. However, the catabolism of GSH conjugates in plants does not yield mercapturic acids as in mammals. The initial GSH conjugation reaction appears to be the biologically significant reaction that detoxifies the herbicide and plays an important role in selectivity.

VI. REFERENCES

1. Kearney, P. C., and Kaufman, D. D., Eds., "Herbicides: Chemistry, Degradation, and Mode of Action." 2nd ed., Vol. I and II, Marcel Dekker, Inc., New York, 1975.

2. Audus, L. J., Ed., "Herbicides: Physiology, Biochemistry, Ecology." 2nd ed., Vol. I and II, Academic Press, New York, 1976.

3. Lay, M. M., Hubbell, J. P., and Casida, J. E., Science 189, 287 (1975).

4. Boyland, E., and Chasseaud, L. F., Adv. Enzymol. 32, 173 (1969).

5. Hutson, D. H., in "Bound and Conjugated Pesticide Residues," (D. D. Kaufman, G. G. Still, G. D. Paulson, and S. K. Bandal, Eds.), ACS Symposium Series 29, p. 103, American Chemical Society, Washington, D. C., 1976.

6. Frear, D. S., and Swanson, H. R., Phytochem. 9, 2123 (1970).

7. Frear, D. S., and Swanson, H. R., Pest. Biochem. Physiol. 3, 473 (1973).

8. Shimabukuro, R. H., in "Environmental Quality and Safety" (F. Coulston and F. Korte, Eds.), Vol. 4, p. 140, Academic Press, New York, 1975.

9. Lamoureux, G. L., Shimabukuro, R. H., Swanson, H. R., and Frear, D. S., J. Agr. Food Chem. 18, 81 (1970).

10. Lamoureux, G. L., Stafford, L. E., Shimabukuro, R. H., and Zaylskie, R. G., J. Agr. Food Chem. 21, 1020 (1973).

11. Shimabukuro, R. H., Walsh, W. C., Lamoureux, G. L., and Stafford, L. E., J. Agr. Food Chem. 21, 1031 (1973).

12. Dauterman, W. C., and Muecke, W., Pest. Biochem. Physiol. 4, 212 (1974).

13. Bakke, J. E., Larson, J. D., and Price, C. E., J. Agr. Food Chem. 20, 602 (1972).

14. Larsen, G. L., and Bakke, J. E., J. Agr. Food Chem. 23, 388 (1975).

15. Lamoureux, G. L., Stafford, L. E., and Shimabukuro, R. H., J. Agr. Food Chem. 20, 1004 (1972).

16. Hutson, D. H., Hoadley, E. C., Griffiths, M. H., and Donninger, C., J. Agr. Food Chem. 18, 507 (1970).

17. Crayford, J. V., and Hutson, D. H., Pest. Biochem. Physiol. 2, 295 (1972).

18. Beynon, K. I., Stoydin, G., and Wright, A. N., Pestic. Sci. 3, 379 (1972).

19. Thompson, R. P., Ph.D. Thesis, University of Illinois, (1974), Xerox University Microfilms, Ann Arbor.

20. Lamoureux, G. L., and Davison, K. E., Pest. Biochem. Physiol. 5, 497 (1975).

21. Shimabukuro, R. H., Lamoureux, G. L., Swanson, H. R., Walsh, W. C., Stafford, L. E., and Frear, D. S., Pest. Biochem. Physiol. 3, 483 (1973).

22. Shimabukuro, R. H., Walsh, W. C., Stolzenberg, G. E., and Olson, P. A., Weed Sci. Soc. Amer. Abstract 196 (1976).

23. Lamoureux, G. L., Stafford, L. E., and Tanaka, F. S., J. Agr. Food Chem. 19, 346 (1971).

24. Shimabukuro, R. H., Walsh, W. C., and Hoerauf, R. A., Pest. Biochem. Physiol. 6, 115 (1976).

25. Lay, M. M., and Casida, J. E., Pest. Biochem. Physiol. 6, 442 (1976).

26. Casida, J. E., Kimmel, E. C., Ohkawa, H., and Ohkawa, R., Pest. Biochem. Physiol. 5, 1 (1975).

27. Chasseaud, L. F., Drug Metabolism Reviews 2, 185 (1974).

28. Shimabukuro, R. H., Lamoureux, G. L., Frear, D. S., and Bakke, J. E., in "Pesticide Terminal Residues," IUPAC (A. S. Tahori, ed.), p. 323, Butterworths, London (1971).

29. Thompson, L., Jr., Weed Sci. 20, 584 (1972).

30. Shimabukuro, R. H., Masteller, V. J., and Walsh, W. C., Weed Sci. 24, 336 (1976).

31. Shimabukuro, R. H., and Swanson, H. R., J. Agr. Food Chem. 17, 199 (1969).

32. Shimabukuro, R. H., Frear, D. S., Swanson, H. R., and Walsh, W. C., Plant Physiol. 47, 10 (1971).

INVOLVEMENT OF GLUTATHIONE AND GLUTATHIONE S-TRANSFERASES IN THE ACTION OF DICHLOROACETAMIDE ANTIDOTES FOR THIOCARBAMATE HERBICIDES

Ming-Muh Lay[1] and John E. Casida

University of California, Berkeley, California

The herbicide antidote R-25788 [$Cl_2CHC(O)N(CH_2CH=CH_2)_2$] temporarily elevates corn glutathione (GSH) content and GSH S-transferase activity thereby facilitating EPTC sulfoxide [$CH_3CH_2S(O)C(O)N(CH_2CH_2CH_3)_2$] metabolism by carbamoylation of GSH. This proposed mode of action in corn is probably also applicable to other thiocarbamates and their sulfoxides and to antidotes related to R-25788. The mechanism by which R-25788 increases the corn GSH content and GSH S-transferase activity is not defined.

I. INTRODUCTION

Herbicide antidotes or protectants or safeners provide an opportunity to extend the use of a specific herbicide or class of herbicides to new areas of crop production. The antidote minimizes or prevents crop injury without adversely affecting herbicidal potency. This requires a specific and temporary effect on the biochemistry or physiology of the crop but not on an analogous system in weeds or mammals. Thus, herbicide antidotes are excellent probes in understanding herbicide mode of action and species selectivity.

Many potentially useful antidote-crop-herbicide combinations have been described in recent years (1-3). Only one such combination will be considered in detail in the present report, i.e. R-25788 which acts in corn to prevent injury from EPTC [$CH_3CH_2SC(O)N(CH_2CH_2CH_3)_2$]. Several suppositions are made in the investigations. First, the antidote alters the ability of corn to detoxify the herbicide. Second,

---

[1]Present address: Biochemistry Department, Mountain View Research Center, Stauffer Chemical Co., P. O. Box 760, Mountain View, Calif. 94042.

mammals and plants metabolize EPTC by similar mechanisms so
findings with mammalian liver enzymes lay the background for
studies with plant enzymes.   Third, the antidote acts in the
same way in young corn seedlings under laboratory conditions
and in various stages of corn development under field conditions.   Fourth, the system altered by the antidote in a
relatively tolerant plant such as corn will not be affected
to the same extent in a very sensitive plant such as oat.
Finally, a mechanism defined for the R-25788-corn-EPTC
combination may be applicable to related antidotes and
possibly to some other crops and herbicides.   Fortunately,
most of these suppositions are easily tested and appear
to be valid as detailed below.

II.   THIOCARBAMATE METABOLISM IN CORN PLANTS AND MAMMALS

Thiocarbamate herbicides are readily biodegraded in
plants and mammals (4) by processes which include: oxidative
attack on the N- and S-alkyl substituents (5-9); sulfoxidation (Eq. 1) (5-7, 10-14); reaction of the sulfoxide with
GSH (Eq. 2) (5,6,10-13); further metabolism of the GSH conjugate (6,12,13).

$$RSC(O)NR_1R_2 + [O] \longrightarrow RS(O)C(O)NR_1R_2$$
$$\text{(Eq. 1)}$$

thiocarbamate   thiocarbamate sulfoxide

$$RS(O)C(O)NR_1R_2 + GSH \longrightarrow GSC(O)NR_1R_2 + [RS(O)H]$$
$$\text{(Eq. 2)}$$

thiocarbamate   GSH   sulfenic
sulfoxide   conjugate   acid

In rats and probably in corn seedlings a large portion of the
EPTC dose is metabolized by sulfoxidation and reaction of the
sulfoxide with GSH (6).   In mammalian liver reaction 1 is
catalyzed by a microsomal oxygenase requiring NADPH as the
cofactor and reaction 2 is catalyzed by GSH S-transferase
enzymes in the soluble fraction.   It is therefore important to
define the involvement of the thiocarbamate sulfoxide metabolite in herbicide action and of plant GSH and GSH S-transferases in metabolism of the sulfoxide.

III.   THIOCARBAMATE SULFOXIDES ARE POTENT HERBICIDES

The observation that thiocarbamates are metabolically

oxidized to the corresponding sulfoxides raises the possibility that this is a bioactivation step required for herbicidal action. This hypothesis is supported by finding that
thiocarbamate sulfoxides (prepared by peracid oxidation of
thiocarbamates) are generally more potent than their precursors in controlling many broadleaf and grass weeds (10,11,
14). The sulfoxides, however, are less injurious to corn
and certain other crops than the corresponding thiocarbamates.
Thus, sulfoxidation often increases the selectivity of thiocarbamates. This change in species specificity on sulfoxidation can be explained if tolerant crops and resistant weeds
detoxify the absorbed sulfoxide as rapidly as it enters the
plant whereas sensitive crops or weeds,lacking this protective
mechanism, absorb and translocate lethal levels of sulfoxide.
Thiocarbamate injury to relatively tolerant crops may result
from absorption and translocation of the thiocarbamate to
cells within the sensitive meristematic region of the stem
(15) where the sulfoxide is formed and rapidly reacts to inhibit growth.

A second line of evidence also implicates the sulfoxides
as the herbicidal agents. Thiocarbamate sulfoxides are much
more effective than the corresponding thiocarbamates as
carbamoylating agents for tissue thiols such as GSH and co-
enzyme A (6,12,13). Possibly the thiocarbamates are converted
to thiocarbamate sulfoxides which then carbamoylate a plant
thiol site important in lipid biosynthesis or other sites
critical for normal growth (12,13). If this is so, a sufficiently active GSH-GSH S-transferase system would prevent
accumulation of the sulfoxide and its disruption of cellular
metabolism.

IV.   R-25788 ENHANCES EPTC SULFOXIDE METABOLISM IN CORN

Several studies with corn indicate that R-25788 increases
the rate of EPTC sulfoxide metabolism. This antidote does
not alter the uptake of [S-ethyl-^{14}C]EPTC from nutrient solutions or its distribution in the plants but it does increase
the amount of volatile ^{14}C products including ^{14}CO$_2$ (16).
The organosoluble ^{14}C products in the stem and leaves of corn
seedlings 8 hr after injecting [S-ethyl-^{14}C]EPTC or -EPTC
sulfoxide (2 μg) into the stem is 2.6-2.9-fold greater in
normal plants than in plants pretreated by stem injection with
R-25788 (100 μg) 24 hr prior to the labeled compounds (13).
In corn seedlings exposed for 24 hr to R-25788 and then for
24 hr to [^{14}C=O]EPTC sulfoxide, the methanol-soluble ^{14}C
products from the roots increase from 9% with no R-25788 to
48% with 30 ppm R-25788 (13). Two of the major metabolites

in this methanol-soluble fraction, the $\underline{S}$-($\underline{N}$,$\underline{N}$-dipropyl-carbamoyl)-derivatives of GSH and cysteine, are 2-5 fold greater in the antidote-treated than in control plants (6). A similar but less dramatic relationship is evident with the aerial portions of corn plants held for 48 hr in solutions of $[^{14}C=O]$EPTC. The leaves and stems contain ~1.4-times more of the GSH conjugate in plants pretreated for 4 days with R-25788 at 1 kg/ha than in control plants but there is little if any difference in the levels of cysteine conjugate and 12 other metabolites (6). Other effects of the antidote include increasing the reaction in the exposed stem section immediately on entering the plant and increasing the conjugate fraction in EPTC sulfoxide-treated plants (6).

V.   R-25788 ELEVATES CORN GSH CONTENT AND GSH $\underline{S}$-TRANSFERASE ACTIVITY

We propose (12,13) that the antidotal action of R-25788 in corn is attributable to elevation of GSH content and GSH $\underline{S}$-transferase activity, thereby facilitating thiocarbamate sulfoxide detoxification so it does not carbamoylate sensitive physiologically-important sites (Eq. 3).

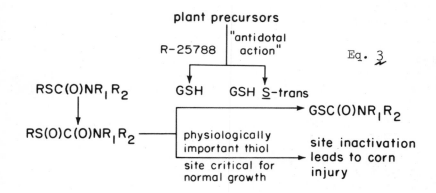

Corn seedlings exposed for 24 hr to increasing levels of R-25788 have a normal root GSH content until a threshhold concentration of 0.03-0.3 ppm is reached and then there is a progressive increase in GSH content up to about 2-fold normal at 30 ppm R-25788 (12,13) (Fig. 1). On varying the exposure time to antidote with slightly older corn seedlings, the increase in root GSH content is evident within 0.5 day and is 2.2- and 2.8-fold within 1 day at 1 and 30 ppm R-25788, respectively. On 2 and 3 days exposure to antidote solution

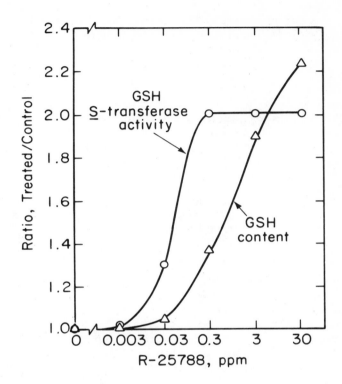

Fig. 1.   Elevation of GSH content and GSH S-transferase activity in roots of corn seedlings exposed for 24-hr to various concentrations of R-25788.

the root GSH content decreases with 1 ppm R-25788, possibly due to antidote depletion under the test conditions, but it increases to 3.6-times normal at 30 ppm R-25788 (13).   The same phenomenon of antidote elevation of GSH content is evident in the leaves, with a ~ 1.5-fold increase on root exposure to 30 ppm R-25788 solution for 24 hr (13).

Three procedures were used to assay the corn root GSH S-transferase activity, with similar results relative to antidote action.   The first studies (12) utilized the soluble fraction of root homogenates (2.5-320 mg fresh tissue weight equivalent), GSH (10 μmoles) and [S-ethyl-^{14}C]EPTC sulfoxide (1 nmole) in pH 6.8 phosphate buffer incubated for 2 hr at 25° C, analyzing unmetabolized [^{14}C]EPTC sulfoxide as the organosoluble ^{14}C products. The second method (13) differed only in using the soluble fraction of homogenates prepared in the presence of polyvinylpolypyrrolidone (to minimize possible inhibitory effects of plant phenolic compounds) as the enzyme source (5-40 mg fresh tissue weight equivalent), a higher substrate level (1 μmole [^{14}C=O]- or [S-ethyl-^{14}C]EPTC sulfoxide) and 40 min incubation. The third procedure (13) was a variation of the second only in analyzing the S-(N,N-dipropylcarbamoyl)-derivative of GSH by thin-layer chromatography. The standard assay (procedure 2 above) provides nearly linear kinetics for EPTC sulfoxide cleavage as a function of enzyme level and time of incubation. The high level of GSH used greatly enhances enzyme activity but the normal level in the cytosol of corn roots and leaves is probably adequate for enzyme action (13). The pH of 6.8 was selected for good enzyme activity with minimal non-enzymatic cleavage of EPTC sulfoxide by reaction with GSH. This enzyme is categorized as a GSH S-transferase since it requires GSH and the product is the carbamoylated derivative of GSH.

It was most interesting to find that R-25788 elevates the GSH S-transferase activity by a magnitude equal to or greater than the increase in GSH content. With 24-hr root exposure, the threshold for antidote effect is about 0.01-0.03 ppm and then activity increases progressively up to 2-times normal or higher until an antidote level of 0.3-3 ppm is reached at which time there is a plateau and no further activity increase (12,13) (Fig. 1). With varying times of exposure to the antidote, the increases in root GSH S-transferase activity parallel those in GSH content, being detectable within 0.5 day and being more dramatic and more persisat 30 than at 1 ppm R-25788 (13). The leaf enzyme is also increased in activity by 2-3 fold similar to the root enzyme (13).

The antidote R-25788 acting in corn performs the remarkable feat of elevating both the GSH content and GSH S-transferase activity thereby providing the cofactor and enzyme for detoxification of EPTC sulfoxide. The antidote does not act directly on the transferase since under in vitro conditions R-25788 at 10^{-5}-10^{-7}M does not alter the enzyme activity (13). The antidote either acts via a metabolite (17) not present in these in vitro assays or more likely by stimulating the synthesis of GSH and GSH S-transferase within the plant itself. The difference in root transferase activity between control and antidote-treated plants is retained on purification of the enzymes by ammonium sulfate fractionation and chromatography on Sephadex G-25 indicating that the

enhancement of enzyme activity is probably due to modifica-
tion within the protein components rather than to the presence of a soluble, low molecular weight activator (in antidote-treated plants) or inhibitor (in control plants) (13).

VI. CORRELATION OF ANTIDOTAL POTENCY FOR COMPOUNDS RELATED TO
     R-25788 WITH INCREASED ROOT GSH CONTENT AND GSH S-
     TRANSFERASE ACTIVITY

There is a reasonably good correlation between the
effectiveness of a series of 29 N,N-dialkyl-2,2-dichloro-
acetamides and related compounds as antidotes for EPTC injury
in corn and their potency, in studies with the roots of corn
seedlings as above, for increasing GSH content and GSH S-
transferase activity (13) (Table 1).

TABLE 1

*Relationship of Antidotal Activity in Preventing Corn Injury
from EPTC to Antidote Effectiveness in Elevating the GSH
Content and GSH S-Transferase Activity of the Roots
of Corn Seedlings*

| Antidotal activity[*] | Number of compounds | Ratio, treated/control ($\pm$ S.E.) | |
		GSH content	GSH S-transferase activity
Superior	4	1.78±0.20	2.25±0.35
Good	5	1.33±0.21	1.69±0.43
Moderate	9	1.22±0.18	1.58±0.30
Little or none	11	1.19±0.12	1.36±0.25

*Compounds applied as seed treatments at about 0.5% by weight
prior to planting in soil in which EPTC was incorporated at
6.7 kg/ha, with ratings of symptoms about two weeks later.

This study strongly suggests that a large number of chemical-
ly-related antidotes act by the same mechanism in alleviating
EPTC injury to corn.

VII. FINDINGS ON VARIOUS ANTIDOTE-ORGANISM-HERBICIDE
      COMBINATIONS

In oat seedlings, which are highly susceptible to growth
inhibition by thiocarbamates and particularly their sulfoxides
(10), R-25788 even at 30 ppm does not elevate the root GSH
content or GSH S-transferase activity (12). This is in marked

contrast to the findings with corn seedlings noted above (Fig. 1). Oat roots contain normal GSH levels only ~ 30% of those in corn roots and no detectable transferase activity, with the assay method utilized (procedure 1 above). It is possible that the antidote elevates the transferase level only in plant species that already have a moderate activity of this enzyme.

The chloroacetamide herbicide CDAA [$ClCH_2C(O)N(CH_2CH=CH_2)_2$] significantly elevates the GSH content and GSH S-transferase activity in the roots of treated corn seedlings whereas the dithiocarbamate herbicide sulfallate [$CH_2=C(Cl)CH_2SC(S)N-(CH_2CH_3)_2$] and the dichloroacetamide antibiotic chloramphenicol are inactive in these assays (13). Of greater interest is the finding that the herbicide antidote naphthalic anhydride does not alter the corn root GSH level or GSH S-transferase activity suggesting that the mechanism of antidotal action of naphthalic anhydride differs from that of R-25788 and related compounds.

The enhanced GSH S-transferase activity of corn roots exposed to R-25788 is evident not only with EPTC sulfoxide as the substrate but also with butylate sulfoxide [$CH_3CH_2S(O)-C(O)N[CH_2CH(CH_3)_2]_2$] (12). However, R-25788 does not alter the recovery of organosoluble products from pebulate [$CH_3CH_2-SC(O)N(CH_2CH_3)CH_2CH_2CH_3$] metabolism in the aerial portions of corn seedlings (13).

Studies with mice indicate that the liver GSH content is not altered 0.25-3 hr after intraperitoneal administration of R-25788 at 1.5 mmole/kg (13). As expected, EPTC and EPTC sulfoxide under the same conditions temporarily lower the liver GSH content, presumably due to carbamoylation of the GSH by the sulfoxide. This suggests that the mechanism by which the antidote temporarily alters the biochemistry of corn plants is not applicable to or is much less prominent in mammalian systems.

VIII. COMMENTS ON OTHER OBSERVATIONS RELATIVE TO R-25788
       ACTION

R-25788 reverses several effects of EPTC including the following: inhibition of fatty acid synthesis in spinach chloroplasts; inhibition of fatty acid synthesis and olelyl desaturase and induction of betacyanin efflux in "aged" red beet root discs; inhibition of polyphenol oxidase activity (18). Naphthalic anhydride also reverses the effects of EPTC in these systems (18). In light of the discussions above, it would be interesting to make these studies by

interacting the antidotes with thiocarbamate sulfoxides rather
than with thiocarbamates.

The antidotal activity of R-25788 extends to a variety
of herbicides (3,19), many of which are metabolized by conjugation with GSH (20). This could result from the antidote
elevating the plant GSH content. It might also involve enhanced GSH S-transferase activity. This points out the
importance of a better understanding of the number, localiza-
tion, substrate specificity and species distribution of plant
GSH S-transferases.

It is interesting to note that with various antidotes
related to R-25788 and at various times after antidote treatment the elevation in corn root GSH content is almost always
associated with increased GSH S-transferase activity. Although it is intriguing to speculate that the antidote serves
as an enzyme inducer, there is no direct evidence to support
this hypothesis. Nevertheless, the association of increased
GSH content and transferase activity suggests the possibility
that they are regulated by the same control mechanism. The
limiting step in biosynthesis of GSH in corn is not established nor is it known if this is the step undergoing antidote-
induced enhancement. In fact, it is not clear whether the
elevated GSH content is due to an increase in de novo GSH
synthesis or to a decrease in GSH destruction (by cleavage,
oxidation or other mechanisms). Attention should be given
to the enzymes involved in GSH synthesis in corn and other
plants and the relationship of these enzymes to GSH S-
transferases using the antidotes as biochemical probes.

IX.   ACKNOWLEDGMENT

Study supported in part by U. S. Public Health Service
Grant 5 P01 ES00049.

X.   REFERENCES

1.   Hoffmann, O. L., in "Chemistry and Action of Herbicide
     Antidotes" (Pallos, F. M. and J. E. Casida, Eds.), p.
     1 . Academic Press, New York, 1978.
2.   Pallos, F. M., M. E. Brokke, and D. R. Arneklev, U. S.
     Patent 4,021,224 (1977).
3.   Blair, A.M., C. Parker, and L. Kasasian, PANS 22, 65
     (1976).
4.   Fang, S. C., in "Herbicides. Chemistry, Degradation, and
     Mode of Action" (Kearney, P. C. and D. D. Kaufman,
     Eds.), 2nd Ed., Vol. 1, p. 323. Marcel Dekker, New York,
     1975.

5. Casida, J. E., E. C. Kimmel, H. Ohkawa, and R. Ohkawa, Pesticide Biochem. Physiol. 5, 1(1975).
6. Hubbell, J. P., and J. E. Casida, J. Agr. Food Chem. 25, 404 (1977).
7. Chen, Y. S., and J. E. Casida, unpublished results (1977).
8. Ishikawa, K., I. Okuda, and S. Kuwatsuka, Agr. Biol. Chem. 37, 165 (1973).
9. Ishikawa, K., Y. Nakamura, and S. Kuwatsuka, J. Pesticide Sci. 1, 49 (1976).
10. Casida, J. E., R. A. Gray, and H. Tilles, Science 184, 573 (1974).
11. Casida, J. E., E. C. Kimmel, M. Lay, H. Ohkawa, J. E. Rodebush, R. A. Gray, C. K. Tseng, and H. Tilles, Environ. Qual. Safety Suppl. Vol. 3, 675 (1975).
12. Lay, M.-M., J. P. Hubbell, and J. E. Casida, Science 189, 287 (1975).
13. Lay, M.-M., and J. E. Casida, Pesticide Biochem. Physiol. 6, 442 (1976).
14. Santi, R., and F. Gozzo, J. Agr. Food Chem. 24, 1229 (1976).
15. Gray, R. A., and G. K. Joo, in "Chemistry and Action of Herbicide Antidotes" (Pallos, F. M. and J. E. Casida, Eds.), p. 67. Academic Press, New York, 1978.
16. Chang, F.-Y., G. R. Stephenson, and J. D. Bandeen, J. Agr. Food Chem. 22, 245 (1974).
17. Miaullis, J. B., V. M. Thomas, R. A. Gray, J. J. Murphy, and R. M. Hollingworth, in "Chemistry and Action of Herbicide Antidotes" (Pallos, F. M. and J. E. Casida, Eds.), p. 109. Academic Press, New York, 1978.
18. Wilkinson, R. E., in "Chemistry and Action of Herbicide Antidotes" (Pallos, F. M. and Casida, J. E., Eds.), p. 85. Academic Press, New York, 1978.
19. Stephenson, G. R., and F. Y. Chang, in "Chemistry and Action of Herbicide Antidotes" (Pallos, F. M. and J. E. Casida, Eds.), p. 35. Academic Press, New York, 1978.
20. Shimabukuro, R. H., G. L. Lamoureux, and D. S. Frear, in "Chemistry and Action of Herbicide Antidotes" (Pallos, F. M. and J. E. Casida, Eds.), p. 133. Academic Press, New York, 1978.

Part VI
Summary

HERBICIDE ANTIDOTES: PROGRESS AND PROSPECTS

John E. Casida
University of California, Berkeley, California

The herbicide antidote concept was introduced by Hoffmann in 1962. Within 10 years, this new approach to enhanced herbicide selectivity had already become an accepted agricultural practice. Corn injury from thiocarbamate herbicides is minimized or avoided by seed treatment with 1,8-naphthalic anhydride or soil application of N,N-diallyl-2,2-dichloroacetamide. Optimization of antidotes for other crop/herbicide combinations has proved more difficult. However, progress is being made in finding safeners for many commercial and candidate herbicides that are limited in their use areas by crop injury. The present antidotes appear to act by increasing herbicide detoxification, but there are also other proposed mechanisms of action for these protectants and possible actions for future antidotes including involvement of growth regulatory effects. Herbicide antidotes serve as new biochemical probes for use, in conjunction with herbicides of diverse modes of action, in expanding the knowledge of comparative biochemistry and physiology of crops and weeds.

I. INTRODUCTION

After nearly four decades of rapid advances in herbicide discovery and development, we have almost reached a steady state situation in chemical weed control. Relatively few new compounds are entering commercial use, new classes of herbicides are discovered less frequently than in the past, and some herbicides are being discontinued because of unfavorable toxicological properties. Patents have expired or will soon expire on most of the current major herbicide chemicals. Increasing development investments and risks, particularly relating to comparative cost/effectiveness and expanded toxicological evaluations, have prompted a

rethinking of objectives. One target of this refocusing has
been on improved and expanded uses of current herbicides with
the goal of enhanced selectivity. This can be accomplished
to some extent by modifications in the timing and placement
of herbicide applications. It is now evident that suitable
protectants or antidotes for crop injury provide excellent
opportunities for improvements in selective weed control.

One unexpected observation and two large and independent
testing programs resulted in the present optimism for antidote research. Hoffmann in 1947 observed that 2,4,6-trichloro-
phenoxyacetic acid counteracts the growth regulatory effects
of 2,4-dichlorophenoxyacetic acid on tomatoes (1). More
significantly, he recognized from this the possibility of
using non-herbicidal compounds to reduce crop injury from
moderately selective herbicides. By 1962 Hoffmann clearly
established and introduced the concept of herbicide antidotes
for crops (1). Major screening programs soon led to the
development of two antidotes for thiocarbamate herbicide injury to corn. 1,8-Naphthalic anhydride for use as a seed
treatment was announced by Gulf Oil Chemicals Co. in 1969
(1) and N,N-diallyl-2,2-dichloroacetamide for use in soil
applications was reported by Stauffer Chemical Co. in 1972
(2). Herbicide antidotes were established as an important
advance in agriculture within 10 years after recognition of
their possible utility.

II.   GOALS IN FUTURE RESEARCH

Each of the commercial herbicides is used at dosages
and in situations where adequate crop safety is usually
achieved. Improved selectivity through antidotes can permit
higher use levels of current herbicides, thereby controlling
weeds only marginally affected at normal rates. It can also
allow applications on crops that would otherwise be damaged,
permitting expanded uses of the present herbicides. Finally,
many very potent herbicide chemicals are known that are not
suitable for use because of inadequate selectivity, i.e.
crop injury. It may be possible to realize the benefits of
these candidate herbicides by combining them with appropriate
antidotes.

The approaches to discovery and optimization of herbicide antidotes are well established (1). The areas of major
needs for safeners in specific crop/herbicide combinations
are evident to weed scientists and herbicide specialists
worldwide. It is clear that antidotes can also reduce the
potency of herbicides to target weeds (3). Thus, the search
for new antidotes must concentrate on the most significant

crop/herbicide/weed combinations. The specificity of antidote
action increases the difficulties in discovering new protectants but it also challenges the imagination and creativity of
herbicide chemists and biochemists.

III.   ANTIDOTE MODE OF ACTION

Scientific advances are often most rapid with a critical feedback relationship between fundamental and applied
programs. New classes of herbicides and antidotes provide
the tools for probing new areas of plant biochemistry. An
understanding of these actions suggests ways and means to
further improve the probes as fundamental tools and practical
weapons in weed control.

There are several hypotheses for the mode of action of the
current antidotes, in fact most or all of the possible actions
have either been tested or proposed. The safeners might react directly with the herbicide in the soil or the crop.
Reactions in the soil would reduce the herbicidal activity.
Increased herbicide degradation by reaction with protectants
in plants would require selective uptake of the antidote by
the crop but not by weeds, perhaps supplementing similar
selectivity in herbicide uptake. In theory, the protectants
can alter the conformation or significance of the physiologically-important herbicide target site, compete with the herbicide in binding to or reaction with this site, or accele-
rate release of the herbicide from this critical locus.
At present this can be fruitfully examined only with the
relatively few herbicides for which a specific target site
is known. Antidotes can potentially inhibit or retard bioactivation of the herbicide in the crop, a proposal not yet
subjected to specific biochemical test. They can enhance
detoxification mechanisms, the glutathione-glutathione S-
transferase system being the current focus with dichloroacetamide safeners. Antidotes for mammalian poisons act by
many of the mechanisms indicated above and they are all
possible within the large variety of plant species.

IV.   CONCLUDING COMMENTS

Chance observations and specific testing programs
have led to herbicide synergists as well as herbicide antidotes. The propanil herbicide/methylcarbamate insecticide
combination is perhaps the most interesting in that it involves inhibition by the methylcarbamate of the herbicide-
detoxifying arylacylamidases in rice and other plants so that
crop injury and herbicidal potency are enhanced (4). It is
often found with insecticide/synergist combinations that the

synergist increases the potency of the insecticide not only to insects but also to mammals (5). Care must be taken to ascertain that the addition of an antidote or synergist to a herbicide does not create new problems by alterations in mammalian toxicity or environmental impact. On the other-hand, it is conceivable that advances in herbicide antidotes for protecting crops might provide an insight into anti-dotes for the same herbicides in mammals. It is therefore appropriate to conduct the tests and biochemical studies with both plant and mammalian systems.

The present herbicide safeners appear to be of very low toxicity to mammals. They act with moderate to high speci-ficity in only certain plants. The marvelous diversity of biochemical mechanisms within closely related organisms once again serves as the focus for new knowledge with potential rapid application. Herbicide antidotes provide new approach-es in understanding the intricacies of comparative biochemis-try and in increasing the efficiency of food production.

V. REFERENCES

1. Hoffmann, O. L., in "Chemistry and Action of Herbicide Antidotes" (Pallos, F. M. and J. E. Casida, Eds.), p. 1 . Academic Press, New York, 1978.
2. Pallos, F. M., R. A. Gray, D. R. Arneklev, and M. E. Brokke, in "Chemistry and Action of Herbicide Antidotes" (Pallos, F. M. and J. E. Casida, Eds.), p. 15. Academic Press, New York, 1978.
3. Stephenson, G. R., and F. Y. Chang, in "Chemistry and Action of Herbicide Antidotes" (Pallos, F. M. and J. E. Casida, Eds.), p. 35. Academic Press, New York, 1978.
4. Still, G. G., and R. A. Herrett, in "Herbicides: Chemistry, Degradation, and Mode of Action" (Kearney, P. C. and D. D. Kaufman, Eds.), 2nd Ed., Vol. 2, p. 609. Marcel Dekker, New York, 1976.
5. Casida, J. E., J. Agr. Food Chem. 18, 753 (1970).

Index